THE IMMEASURABLE WILDS

THE IMMEASURABLE
Wilds

TRAVELLERS TO THE FAR
NORTH OF SCOTLAND

1600-1900

ALASTAIR MITCHELL

Whittles Publishing

Whittles Publishing Ltd.,
Dunbeath,
Caithness, KW6 6EG,
Scotland, UK

www.whittlespublishing.com

© 2022 Alastair Mitchell

ISBN 978-184995-492-1

Printed by Short Run Press Ltd.

CONTENTS

PERMISSIONS

Most of the images reproduced in this book are from the author's collection. Thanks are due to the National Library of Scotland for making their maps available. I have used their image of John Cowley's *A Display of the Coasting Lines of Six Several Maps of North Britain* [EMS.b.6.5(1)] and Timothy Pont's sketch, a detail from the *Durness and Tongue Map* Adv. MS.70.2.9 (Pont 1). The National Library of Scotland has also granted permission to quote from MS.9233 *Journal of a Few Days from Home*; MS 2507 and MS 2508 (the Robertson Journals); and also NLS, Acc 11529, the *Diary of the Reverend MacDonald of Durness*. The Murchison Correspondence and papers are held variously at the British Geological Survey, Edinburgh University, Elgin Museum and the Geological Society of London. Geikie's field notebooks can be found at the Haslemere Educational Museum. The British Geological Survey gave me permission to publish my copy of the 1913 sheet 92 map EA20/047, and the British Library Board sanctioned the use of the Great Map image; O maps cc.5.a.441 34/3d. Thanks are due to the National Galleries of Scotland for permission to use two images from the frieze that surrounds the foyer of the National Portrait Gallery (Murchison and Telford), and for the use of the Portrait of the Morton family by Jeremiah Davison (PG2233. National Galleries of Scotland purchased 1974) and the Paul Sandby sketch of Road Builders (D2343 National Galleries of Scotland. The William Findlay Watson Bequest 188).

Permission was granted by Boydell & Brewer to quote from the interesting account *To the Highlands in 1786* which is edited by Norman Scarfe. Much of the information regarding Timothy Pont comes from the work of Jeffrey C. Stone, and can be found in his book *The Pont Manuscript Maps of Scotland*. Finally, no one who writes about the Highlands Controversy can ignore David Oldroyd's superbly researched book of that title, and my debt to him increases when I acknowledge that my discovery of his work led in part to my desire to write *The Immeasurable Wilds*.

ACKNOWLEDGEMENTS

Putting this book into some sort of shape was a solitary effort. Various people then looked at what I had written, and to them I wish to express my thanks for making the whole a good deal more presentable. Stephen Young studied each chapter with great care and attention to detail, and offered suggestions that made the text very much more accurate and logical. Peter Harrison made sure there were no geological blunders. His field trips under the auspices of the North-West Highlands Geopark were both fun and illuminating, even if they haven't made me into a serious geologist. Chris McNeill has spent time looking at my collection of untitled images, trying to work out the geographical locations of the places they depict. His knowledge of Scotland has been of great help to me, and he has allowed me to use two of his photographs. Jacob King assisted with some of the Gaelic I came across. I offer particular thanks to the map department of the National Library of Scotland, not least for putting their collection online, and making it so available to everyone. In particular, Christopher Fleet was very helpful in the early stages of the project. My long-suffering wife, Veronica, deserves special thanks for reading and re-reading my literary efforts, and for knocking them into a format that I could never have achieved on my own. Various friends have offered specific advice and encouragement from time to time, and I end with much gratitude to Whittles Publishing, who agreed to take on the task of getting this book into print. To all, I offer heartfelt thanks.

Loch Eriboll, looking south, with Ard Neakie in the foreground. An early 20th century photograph, taken when the lime quarry was still being worked.

FOREWORD

I am standing overlooking Loch Eriboll. To my right is the small isthmus of Ard Neakie, on which stand the old lime kilns built in 1870, and the house that still bears the worn remains of the Sutherland coat-of-arms over the door. Beyond, towering over the western shore of the loch, the peaks of Ben Spionnaidh and Cranstackie rise up, leading the eye down towards the southern end. There in the distance can be seen the white quartz-topped summits of Foinavon, and further still, Arkle; names made famous in the 1960s, not by mountains, but by horses. The Duchess of Westminster owned both the land and the thoroughbreds, one of whom – Foinavon – won the Grand National in 1967 at odds of 100/1, while the other enjoyed an acclaimed racing career. Arkle the horse may have outstripped Foinavon in racing glory, but Foinavon, the mountain, exceeds Arkle in both height and majesty.

To my left, a smaller ridge crowned by Ben Arnaboll is just high enough to obscure the view of Ben Hope, the most northerly 'Munro', which stands on the eastern side of the parallel valley. If I lower my eyes from the skyline, I see land that was farmed by the Clarke dynasty, during the controversial era of the clearances. The family has farmed in Sutherland for over 200 years. Another Clark, Alan, had a house here too, but there was no direct connection between the families. His diaries still bring to life the years of the Thatcher government, and he wrote of 'the Eriboll magic' (Diary, 3 July 1999) on what he knew was his last visit to the area, when suffering from terminal cancer.

The loch itself is deep. Blaeu's mid-seventeenth century maps of Scotland, based on those of the surveyor Pont, recorded a hole to a depth of 180 fathoms at the southern end. The sheltered waters encouraged visits from the navy, especially during the Second World War. The sailors called it Loch 'Orrible, and to pass the time, spelt out the names of their ships – Hood, Amethyst – on the hillside above the village of Laid, using the plentiful bright quartzite rock.

Today, the sky is cloudy, with occasional flecks of blue – 'enough to fill a Highlander's breeks [trousers]' as my mother would have said. The water glistens with a silvery sheen as dark clouds approach from the west, the mist settling on the hills. The silence is disturbed only by the increasing wind. I sense the power of the landscape, almost as if I can now feel the movement of the rocks beneath my feet inching their way inexorably westward, as once they did.

I first went up to Sutherland in the 1960s. A family friend had taken Melness House on the Kyle of Tongue for two weeks in August, and we were all invited to join her. Both my grandparents lived in Scotland so we were used to the trek up from Kent, where I was born, but none of us had ever been so far north before. My parents knew the Highlands well, but I remember their astonishment and delight at the sight of the vast expanse of unpopulated moorland that stretches east and west on either side of the road that leads from Lairg to Tongue: 40 miles of single-track road, with passing places. The novelty of such a road, the A836, no less, was a wonder to a child like myself. I peered out from the back seat: would the oncoming vehicle say thank-you when we had stopped for it? We passed through Crask, a village that merited a sign both on the way in and out, yet consisted of only one house, an inn; then Altnaharra. These were the only markers on the route – all so exciting, yet also frustrating for a 12-year-old who just wanted to get there.

When we did arrive, we found a magnificent Georgian farmhouse, with wonderful stone floors in the kitchen, and a walled garden that, even in its untended state, produced endless raspberries for us during our stay. The view across the Kyle to Ben Loyal was unforgettable, and below the house, seals slumped onto the sand banks that were exposed as the tide went out. I had never seen so much sand, vast stretches of beach with no one there. 'Cow beach', because cows came down to it in the evening. 'Wreck beach', because there was an old wreck marooned on it, slowly rotting away. 'Stone beach', because it was full of rounded pebbles of an alabaster white. And a sea of such clarity, and, on a good day, such a variety of blues and turquoises, a sea which, to a 12-year-old, really wasn't *that* cold!

Sutherland has not changed a great deal since those days. For most visitors to Scotland, the north ends at Inverness, from where they head west to the Kyle of Lochalsh and Skye, or back south along Loch Ness. Tell these people that there is still nearly 100 miles of land to the north, and they look at you in amazement. The NC500, a suggested coastal route launched in 2015 has brought more people into the area, which can only be a good thing, though those who treat it as a racetrack have done no one any favours.

I think that, apart from a love of this quite exceptional part of Britain, the idea for this book emerged from two discoveries. First, having an interest in old maps, I came across a copy of the Reverend Alexander Bryce's 1744 map of the north coast. I was intrigued by the thought of what lay behind it, and was delighted to find what an interesting man the good reverend was. Secondly, I picked up a copy of David Oldroyd's *The Highlands Controversy* in a bargain bookshop. I had never heard mention of the affair, but here was a

*Melness House and farm. An early 20th century photograph,
with the Kyle of Tongue behind, and Ben Loyal in the distance.*

book full of photos of places I knew – Loch Eriboll, Ben Arnaboll, Glencoul – and a story of the utmost importance and fascination to me, even if most of the geology went straight over my head.

I then began to wonder who in the past had made the effort to visit these northern regions, and why – particularly when travel was so difficult before the roads were made. My research has introduced me to all sorts of delightful characters, many of them intrepid, determined types. In addition to the better-known writings of travellers like John MacCulloch, I particularly enjoyed the accounts of the Reverend James Hall (though they get a bit thin when he reaches the north coast), and Catherine Sinclair's *Scotland for the Scotch* which was published in 1840. Catherine's father, Sir John Sinclair, editor of the first *Statistical Account of Scotland* was, according to Sir Walter Scott, '…that Great Caledonian Bore' (Letter to JW Croker, 5 January 1823, quoted in Trevor-Roper, 2014). His daughter, however, has a disarmingly light sense of humour, as displayed on the occasion when, on hearing that it was available to rent, she tried to gain access to Coul House. She was met at the lodge by a lady resembling one of Macbeth's Witches, who gave her what Catherine terms 'the true meaning of a Coul reception' (Sinclair, 1840).

A scarce photograph, said to be Catherine Sinclair (1800–1864), the author of Scotland for the Scotch.

There are plenty more such anecdotes, though not all have found their way into the book. It is as much a history of the Highlands, as an account of travels and travellers, but it is important always to bear in mind that it is the Highlands seen through the eyes of these travellers: the voice of the Highlander himself is rarely recorded. The journalist and newspaper editor James Browne was so incensed by what he read in these accounts that he published in 1825 a *Critical Examination of Dr MacCulloch's Work on the Highlands and Western Isles of Scotland*, a work of 302 pages of intense invective. He refers to Pennant's 'foppery, pedantry, prejudice and folly' and thinks Dr Johnson's account is 'full of grumbling, saucy, ill-natured observations, the spawn of a mind contracted and illiberal, deeply imbued with prejudice, and incomparably more enamoured of antithesis than truth'. In general, he decries all the 'plodding antiquaries, crazy sentimentalists, silly view hunters, cockney literati, and, worst of all, impudent "Stone Doctors" [e.g. the geologist Dr John MacCulloch and his like] armed with their hammers [who] have successively invaded the unconquered mountains of Caledonia'. The *Dictionary of National Biography* tempers Browne's criticisms, commenting that he was known for the 'excitability of his temperament', and a 'boisterous, blustering mode of expression'. It led at one point to a duel with the editor of the Scotsman newspaper, which both thankfully survived.

My research has included an interest in original images of Scotland. The Victorian sketch (these do not constitute great art, but are often nonetheless of a very high standard) is the nineteenth-century equivalent of the twentieth-century postcard or personal photo. There are plenty that depict the southern Highlands – Loch Lomond, Loch Katrine – but far fewer further north, reflecting the number of people who journeyed to these regions. I do not claim that my research in general has been fully comprehensive. My main interest was in those who had travelled to the far north, but I have sometimes included visitors who struggled further to the south (for travel in Scotland was never easy), some of whom crossed over to the Hebrides. I couldn't bring myself to include much from Burt's *Letters from a Gentleman...*; while it is an important record, he has such a low opinion of the Highlander.

A 19th century anonymous sketch of an artist at work, drawing
the "Brig o' Brachlin" – the Bracklinn Falls near Callander.

In an effort to come to terms with the geology, I have been on two geological weeks, led by the genial Peter Harrison of the North West Highlands UNESCO Global Geopark. I am still no geologist, but hope readers will share my fascination with the nineteenth-century Controversy, that for me is more about people than science. No one has traversed these remote regions in more detail than those nineteenth-century geologists.

On our second field trip, we in triumph found the 'Lapworth Sections', a key geological feature situated near Heilam beside Loch Eriboll. Here we had our lunch, high up with the loch to our left, and the land stretching out to the cliffs of Whiten Head ahead of us to the north. There is a silence in this area that can be matched by few other places in the British Isles these days. Suddenly, a noise of several souped-up engines cut in, and half-a-dozen sporty vehicles appeared below us, clinging to the road that curved to our left. 'Stop! Stop!' cried Peter, holding up his arms as if he were King Canute at the water's edge. 'You are missing the Lapworth Sections!'

It is to be hoped that this book will appeal to all who love this area, and maybe even encourage a few more to slow down, and see what is there as they speed along the NC500.

ONE

A COUNTRY SO TORN AND CONVULSED

Our writers hath hitherto erred in descryving the situation of Sutherland.

(Robert Gordon)

Sutherland: a strange name for the county that lies at the extreme north of the British mainland, far from the rich loam lands of the Borders, from the industry of Glasgow, and the sophistication of Edinburgh. Far, too, from the mountains that lie to the north of Stirling and Perth, the stage on which much of the romance of Scotland was set – Rob Roy country, the Pass of Killiecrankie, Glencoe, the Road to the Isles. Climb Ben Nevis and you will peer in vain as you look to the north for a glimpse of the county. Continue to Inverness, and still 40 miles separate you from the southern border of Sutherland, and 80 miles from the imposing cliffs of the north coast, with the cold Atlantic stretching to the horizon, uninterrupted by land until it reaches the arctic regions.

A remote corner, perhaps, but one that has always been visited and populated. The name Sutherland (Old Norse, Suðrland, Southern Land) is testament to that: the region was the southern point of the Norse earldom of Orkney, established in the late ninth century, which stretched from Shetland to the northern mainland. Even today, Shetlanders consider themselves to be as much Scandinavian as British, and the link from Orkney to the southern mainland is tenuous. Even before the arrival of the Vikings, it was a territory that had been inhabited for a long time. The stone brochs, hill forts, hut and stone circles, tombs and cairns that are to be found across the area form the evidence of that earlier civilisation. They are preserved in surprising numbers because much of the land has remained uncultivated for centuries. The population at that time was living in a very different environment, the climate warmer, and the landscape heavily wooded. Now, as Dr Johnson observed further south, it is a land of 'stone and water. There is, indeed, a little earth above the stone in some places, but a very little; and the stone is always appearing. It is like a man in rags – the naked skin is still peeping out' (Boswell, 1984). John MacCulloch thought it a 'far-extended ocean-like waste of rocks and rudeness' (II, 1824).

The earliest maps, dating from the thirteenth century, describe the area as 'A country marshy and impassable, fit for cattle and shepherds,' and also as 'a mountainous and woody region, producing a people rude and pastoral, by reason of marshes and fens' (thirteenth-century map by Matthew Paris). Another states *Hic Habundant Lupi* – 'Here wolves abound' (anonymous thirteenth-century map). However, according to the account found in Sir Robert Gordon's *Genealogical History of the Earldom of Sutherland*, written in 1630, the area was not simply infested by wolves, but was a veritable pastoral haven. 'The country or province of Sutherland abounds in corn, grass, woods, fruit, beasts, all kinds of wildfowl, deer and roe; all sorts of fish, especially salmon, and all other commodities which are usual in this kingdom of Scotland, or necessary for man.'

At Durness the inhabitants chase the deer into the ocean 'where they do take them in boats.' Gordon tells of four large forests 'delectable for hunting.' They are full of deer, wolves, foxes, wild cats, badgers, squirrels, stoats, weasels, otters, pine-martens, hares, and polecats.

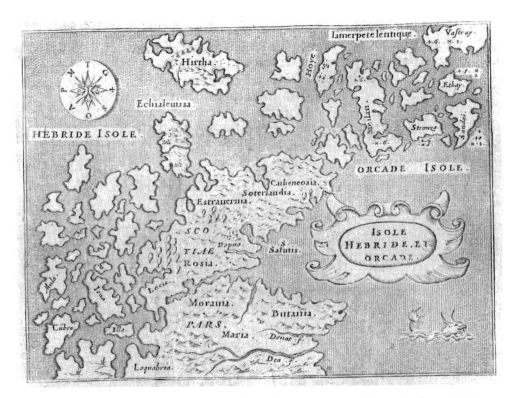

The early maps of the north of Scotland are distinctly approximate. This one by Poracchi shows the Hebrides and Orkneys scattered in a random fashion around the mainland (c.1575).

He then goes on to list 30 types of birds, from plovers and capercaillies to blackbirds and song-thrushes and even the woodpecker, which he calls the 'lair-igigh, or knag, (which is a fowl like unto a paroquet, or parrot, which makes a place for her nest with her beak, in the oak tree)'.

The seas offer all sorts of shellfish and seabirds, and sometimes even huge whales. The county has five major rivers, the Shin offering 'the biggest and largest salmon in the kingdom.' It also has fine pearls, 'some whereof have been sent unto the king's majesty into England, and were accounted of great value.' Other minerals abound – marble, coal (mined at Brora), and, in various parts of the county, iron ore. There was said to be silver too. The Provost of Aberdeen, Sir Thomas Menzies, found some and took samples to London which 'were found to be very rich.' Unfortunately, he died on his way back north without disclosing where he had found it.

The principal town in Sutherland was Dornagh (Dornoch), which was surrounded by green fields on which the locals could be found practising archery, riding and even golf. It was first recorded there in 1619. Corn was grown in plenty in the surrounding countryside, whilst at Dunrobin a little to the north, were found all sorts of fruit, herbs, and flowers, and even saffron, and tobacco.

However, all this abundance was to be found only in the south-eastern corner of the county. Cultivation in the north and west was always a struggle. Here, fish was plentiful, but Gordon voices a criticism that will become familiar: 'If the inhabitants were industrious, they might gain much by these fish; but the people of that country are so far naturally given to idleness that they cannot apply themselves to labour … there is no doubt but that country might be much bettered by hard-working and painstaking inhabitants.'

The county was assessed in 1655 for Oliver Cromwell by Thomas Tucker, who took a rather cooler view of the area's potential than Sir Robert Gordon. He reported that the rivers were 'too big for any trade that is or may be expected in these parts.' He found only two harbours in Ross-shire, again both on the east coast, one at Tain, the other, 'one of the most delicate harbours reputed in all Europe,' at Cromarty. Even so, the only trade was 'a little salt to serve the country, and … may be a small barque once in a year from Leith, to fetch sheep, which are brought down thither from the hills.' He found that it had never been considered necessary to appoint a customs official to either county. As for the regions to the west – Strathnaver, Assynt, and the Western Isles – he was even more dismissive. 'Places mangled with many arms of the Western Sea….and destitute of all trade, being a country stored with cattle, craggy hills and rocks, and populated with the ancient Scots or wild Irish, whose garb and language they do still retain amongst them' (quoted in Hume Brown, 1891).

Access to these areas was always difficult, and much that was reported was often based on hearsay, rather than direct experience. It was a land of myth and rumour. Thus Jorevin de Rocheford, travelling in 1661, writes 'It is said, that there are certain provinces on that side of the country [the north west], where the men are truly savage, and have neither law nor religion, and support a miserable existence by what they can catch' (quoted in Hume Brown, 1891).

"Cromarty Harbour" by J.C. Nattes. From Scotia Depicta *by James Fritter, 1819.*

An even worse assessment is to be found in Thomas Kirke's *A Modern Account of Scotland by an English Gentleman* (1679). Referring to Highlanders, he states, 'The people are proud, arrogant, vain-glorious boasters, bloody, barbarous, and inhuman butchers. Deception and theft is in perfection among them, and they are perfect English haters.' He comments on the cruelty they were said to show their animals, describing in grisly detail how they slowly feast on their cattle, 'till they have mangled her all to pieces … such horrible cruelty as can scarce be paralleled in the whole world' (quoted in Hume Brown, 1891).

Richard Franck in his *Northern Memoirs* was keen to imply that he had traversed the whole of Scotland. The book, which was written in 1658, takes the form of a dialogue between Theophilus and Arnoldus, whose principal delight was the sport of fishing. Franck may well have visited Dornoch and Dunrobin, but when it comes to describing the more inaccessible parts inland, he resorts to the usual hearsay. Like Kirke, he delights in describing how the population kill and devour their cattle 'since few or none amongst them hitherto have as yet understood any better rules or methods of eating.'

On the subject of vermin, he doubts the 'absurd opinion' of the locals that 'the earth in Ross has an antipathy to rats', though he admits that when in the county, he never saw one. Gordon, too, ascribes this characteristic to the earth of Sutherland, adding that there are lots of rats in the adjacent county of Caithness. It appears that people from other regions in Scotland as well were so convinced of this property that they would take soil from Ross-shire and spread it on their own fields, hoping it would deter vermin in their area. Whilst Franck was sceptical of that claim, he seemed prepared to accept the rather more remarkable assertion that in the Orkneys barnacle geese grew on fir trees. Those that dropped off and fell onto dry land perished, but those that fell into water were destined to live.

Such were the tales on which the general perception of the far north of Scotland was based – hardly conducive to encourage travel and exploration. Even as early as the fifteenth and sixteenth centuries, there were regular visitors to Scotland from Europe, though none seem to have ventured deep into the Highlands. Jean de Beaugue, sent by King Henry II of France in 1548 to assist the Scots in their struggle against the invading English, was unusual in getting as far north as Aberdeen and Montrose, but he kept well away from the north and west. There were, however, some people whose business required them to venture into those regions: surveyors and mapmakers. One in particular stands out at this time – Timothy Pont. He was said to have surveyed the whole of Scotland on his own – an exceptional achievement in itself – and a number of his original sketches have survived to this day. Even more remarkable is the fact that, after his death, those rough sketches found their way to the most prestigious cartographical publishing house in Europe, that of Joan Blaeu in Amsterdam, and onto the pages of that most lavish of atlases, *Theatrum Orbis Terrarum*. With the publication of Volume V in 1654, Scotland, hitherto an almost unknown country, became at a stroke one of the best mapped areas in Europe. Unlike some cartographers, Blaeu was willing to reveal his sources: on the majority of the county maps can be found the words *Auct. Timothy Pont*. The acknowledgement on the map of the Hebrides is even more explicit: *Lustratae et descriptae Timothy Pont* – 'Surveyed and described by Timothy Pont.'

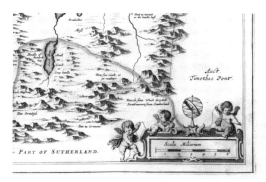

Detail from Blaeu's Map of Strathnaver, taken from his Atlas Theatrum Orbis Terrarum, *1654.*

For all the wealth of detail Pont has left us concerning Scotland, we know little about the man. His father was Robert Pont, an important figure in the Church, who travelled widely on ecclesiastical business, probably taking Timothy with him. We do not know the name of his mother, but she gave birth to two sons and four daughters. It is presumed she died young, as Robert remarried (possibly twice). Timothy was the younger son, but exactly when he was born is unknown – probably sometime between 1562 and 1565. We have no idea who, if anyone, commissioned the survey, nor when it was carried out, though the Dean of Limerick, Denis Campbell, stated in April 1596 that he 'is informed one Pont, who has compassed the whole of Scotland, purposes to set forth a perfect description of that land' (letter to Sir Robert Cecil, 1596, quoted in Chambers, 1857 and Stone, 1989). This suggests that the survey had been completed by then. Timothy married Isobel Blacader, but there is no record of any children. He became Minister of Dunnet in Caithness in 1601, and was 'unhappily surpris'd by death' sometime after 1611, 'to the inestimable loss of his country, when he had well nigh finish'd his papers' (Bishop Nicholson, 1702, quoted in Chambers, 1857 and Stone, 1989). As far as we are aware, he left no will. In spite of all the shadows, Pont remains a fascinating figure.

Various scholars, notably Jeffrey Stone, have pored over his sketches, which have taken some deciphering. These maps differ from one another in style and presentation. Some are crude working sketches, while others are better finished. Pont employs two different handwriting styles – italic, and what is called 'secretary hand' which was quicker to write, but is now harder to read. To complicate things further, at least two other people have written on the maps, adding their own annotations – Robert Gordon of Straloch (of whom more later), and Robert Sibbald, who eventually took possession of the sketches. Pont also uses three different units of measurement when describing distances: miles (the Scottish mile of 2,400 yards), bowshots (200 yards, the range of an arrow), and bowbutts (30 yards, the distance between mounds for archery practice).

Pick your way through the details, however, and a vast amount of information emerges. Much of it is what Sir Robert Gordon later confirmed in his book on Sutherland. In Eddrachillis, there is '[e]xtreem Wilderness… many woulfs in this country'. At Loch Shin, 'the mightiest and largest Salmond in al Scotland', and at Loch Gareron [Loch Inchard], also 'a great fishing'. Laid-Kynnoin [sic] on the other hand provides 'excellent fyn hunting'. In Stranaver [sic], there is the puzzling entry, 'I fand the 2 kynds of mos wyld Berries with ther floures in the head of Korynafairn [sic]', and, more alarmingly, at Loch Stack, 'all heir ar black flies in this wood … seene souking mens blood'. Durness, a more fertile limestone region, is described as 'Gallant country of corn'. Mills, mines, and harbours are depicted, including the coal mines at Brora. At the head of Loch Stack, he states, 'Heir Pearle', and at various places in Strathnaver, 'Heir yron oare', suggesting that there was a thriving iron industry in the region. There was certainly an important area for smelting which was centred on Loch Maree. Sir George Hay, a favourite at the court of James VI who seems to have fallen out of favour, found sanctuary in Ross-shire, and initiated iron production at Letterewe, Talladale, and the Red Smiddy near Poolewe. He employed 'English Miners' (i.e., not Gaelic speakers) from 1607, and as long as the wood was plentiful, the industry thrived. But it was not an unlimited supply; one furnace would consume 120 acres of wood in one year. The evidence of this industry remains today: 'Furnace' is a small place marked on the map south-east of Letterewe, and the *Cladh nan Sasunnach* or 'Englishman's Churchyard' (see colour section, page ii) can be found towards the south-eastern end of the loch. It is thought that a number of these 'English' workers died in an outbreak of smallpox and were buried there.

Some of the earliest biographical detail relating to Pont comes from William Nicholson (1655–1727), who was bishop of Carlisle from 1702. A man noted for his fiery temperament, he was also a keen antiquarian and wrote an *English Historical Library* between 1696 and 1699. He followed this with one on Scotland in 1702, in which he wrote that Pont 'was by nature and education a complete mathematician, and the first projector of a Scotch Atlas.' It is thought that Pont spent some time at St Andrews University studying under William Wellwood, who taught, amongst other subjects 'the making of the cartes universall and particular' (Cunningham, 2001). But Pont's maps are not cartographical in the sense of plotting points with the utmost accuracy, nor are they designed as navigational aids. His aim was to give a general description of the area, a wider impression than pure cartography can provide. Chorographic rather than cartographic, to use the proper term. There is no

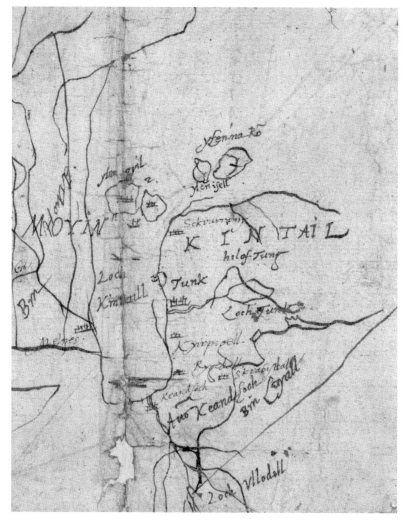

Detail from one of Timothy Pont's sketches of the north coast, showing the Kyle of Tongue (Loch Kintail) and the outline of Ben Loyal. Image courtesy of the National Library of Scotland.

evidence that he employed any triangulation techniques in his surveying: rather a simpler compass traverse using the higher ground as surveying points, and following the rivers. His depiction of mountains is often very particular – Ben Loyal is seen in profile from the shores of Loch Loyal, and An Teallach is portrayed with its recognisable rocky ridge. This is a surveyor who undoubtedly had covered the ground, and who sought to capture all aspects of it.

Bishop Nicholson was insistent that Pont 'personally surveyed all the several counties and isles of the Kingdom'. Robert Gordon of Straloch was even more specific:

> [Pont] travelled on foot right through the whole of this Kingdom, as no one before him had done; he visited all the islands, occupied for the most part by inhabitants hostile and uncivilised, and with a language different from our own; being often stripped, as he told me, by the fierce robbers, and suffering not seldom, all the hardships of the dangerous journey, nevertheless at no time was he overcome by the difficulties, nor was he disheartened (letter to Sir John Scot, 1648, quoted in Stone, 1989).

The reference to robbers contradicts everything one reads about the honesty of this northern area, but later mapmakers found that a stranger seen carefully taking notes was viewed by the locals with suspicion.

Inevitably in such a project there are errors. The accuracy of the north-west coast in Ross-shire and Sutherland is particularly poor. Handa Island is in the wrong place, and the depiction of Loch Crocach bears no relation to its true shape. Lochs Glendhu and Glencoul seem to have been united as one loch, joined to Loch a' Chairn Bhain by a small stream which he calls Alt Glendow. This bears little resemblance to the magnificent scene one looks down onto from the viewpoint above Kylesku.

The biggest error concerns Cape Wrath, which is particularly unfortunate as its legacy affected Scottish cartography for the next 150 years. For some time, there had been confusion as to whether or not Cape Wrath and Faraid Head were one and the same place. Mercator had marked it thus on his large 1564 map *Angliae, Scotiae et Hiberniae nova descriptio*, and from this headland, the coast is plotted as descending in a south-westerly line, thereby slicing off the portion of land that continues along the north coast to Cape Wrath itself. Some maps even show Durness positioned on the west rather than the north coast. Pont had the chance to correct all this, but he did not. It is surprising, for he almost certainly went out onto Faraid Head. Stone thinks he might have stayed at Balnakiel, from where a walk onto the headland would have been an obvious thing for him to do. From this point, looking to the west, the coast with its spectacular cliffs can be seen to continue in a north-westerly, rather than a southerly, direction. One is forced to assume that the difficult terrain in these parts defeated the intrepid surveyor; the various errors suggest that he never explored for himself that part of the coastline.

Scotiae Tabula III, *a detail from Mercator's* Atlas Minor, *1607,*
showing Cape Wrath and Faro Head combined as one headland.

It is not clear who commissioned the work, though Bishop Nicholson describes the laird, advocate, and antiquary, Sir John Scot of Scots-Tarver [Scotstarvit], as the man 'who encouraged Timothy Pont in the survey of the whole kingdom' and later as 'his patron'. To what extent this patronage included financial backing is not clear, but when the work was completed, Pont does not seem to have been under any pressure to pass the sketches on to anyone, and no one came forward to claim them on his death. Indeed, Nicholson states that 'his maps were so carelessly kept by his heirs, that they were in great danger from moths and vermin' (Nicholson, 1702).

The means by which the maps found their way to Blaeu was complicated. Any sketches of so little known a country as Scotland were bound to be of interest for reasons of national security, and there is some suggestion that both King James I and King Charles I showed a desire to have them published. What is certain is that Sir John Scot had access to them. Blaeu knew Scot, having previously published a volume for him in the *Delitiae Poetarum* series, and he approached Sir John seeking information for his intended Atlas of Scotland. Scot realised that the sketches were something of a goldmine, and brought in Robert Gordon to

*King Charles I points to Scotland in
this engraving "Fidei Defensor" by
William Marshall (c.1650).*

*Robert Gordon of Straloch, 'eminent geographer
and antiquary' (Chambers). An engraving after
the portrait by George Jameson.*

edit and amend them so that they could be of use to the Dutch cartographer. Gordon (not
to be confused with Sir Robert Gordon of Dunrobin) was an antiquarian and cartographer
of note, living at Straloch in the southern Highlands. Judging by his statement, 'as he told
me', he knew Pont personally, which would not have been surprising given his interests.
Gordon's annotations on the sketches are part of this editing in preparation for publication.
The maps were then offered to Blaeu, and many details from them found their way into the
atlas. One can only imagine the delight and pride Pont would have felt had he been alive to
see the lavish volume, published in 1654, with its superb engraving and magnificent hand
colouring.

Pont's survey must be recognised as an exceptional achievement. His work is matched only
by the later surveys of General Roy, and the Ordnance Survey, both of whom had at their
disposal teams of soldiers. Many of the maps that followed in the seventeenth and early
eighteenth centuries are simply redesigned copies of earlier ones, notably those by Blaeu,
John Speed, and Mercator. A surveyor called John Adair announced his intention of 'the

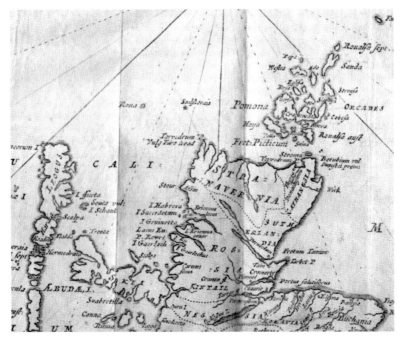

A detail from John Adair's map Nova Scotia, *from the 1727 edition of George Buchanan's* History. *Cape Wrath is still missing and the shape and position of all the islands and rocks are still very approximate.*

Surveying of all Shires of Scotland and making new mapps of it' (1681, quoted in Moir, 1973), but his efforts were hampered by financial difficulties. In 1696 he appealed to the Privy Council, stating, 'it is thought very hard, that John Adair should spend so much of his time, undergo so much travel and hardship, yea, and frequently run the hazard of his life, and only get back his bare disbursements, and that not without difficulty' (Moir, 1973). For 30 years he struggled in the field, apparently more interested in the surveys than in getting his work published, with the result that his impact on the mapping of Scotland was less than it might have been. At the time of his death in 1718, maps were still being published with Cape Wrath sliced off as if it did not exist, in spite of the fact that shipping was rounding this most prominent headland on a regular basis.

A map published by John Cowley in 1734 highlighted the problems that beset the accuracy of Scottish cartography in a most remarkable way. *A Display of the Coasting Lines of Six Several Maps of North Britain* imposes the outlines of the country one on top of another, as drawn by various cartographers: John Adair, Hermann Moll, Robert Gordon, Nicholas Sanson, John Senex, and Charles Inselin. The differences are alarming and speak for themselves. However, within ten years, a significant survey along the north coast had been completed; this was to make a huge contribution to Scottish cartography. To understand

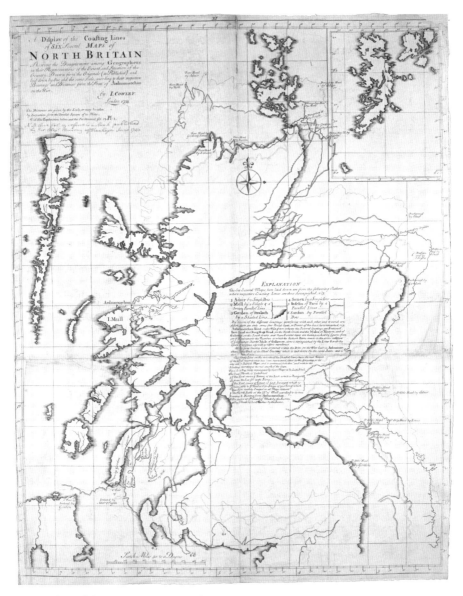

A Display of the Coasting Lines of Six Several Maps of North Britain. *John Cowley's 1734 map compares versions of the coastline of Scotland as depicted by six cartographers. Image courtesy of the National Galleries of Scotland.*

how this came about, it will be necessary to leave the Highlands for a while, and spend some time in the capital of the country, Edinburgh, the 'Athens of the North.'

The Scottish Enlightenment was a remarkable movement. During the eighteenth century, men like David Hume and Adam Smith were at the forefront of intellectual thinking, not just in Scotland and Great Britain, but in Europe as well. At no time was the difference between the Lowlands and the Highlands more marked in Scotland: while the north was isolated and inward-looking, the south of the country was an open door to Europe, leading the way in philosophical and scientific thought.

Colin MacLaurin was a typical figure in this movement. Born in 1698, he was drawn to mathematics at an early age, had mastered Euclid by the time he was 12, and had an M.A. degree at the age of 15. At a time when the theories of Newton were barely understood, he chose as his special subject the power of gravity. He was appointed to the Chair of Mathematics at Aberdeen University at the age of 19, and was admitted to the Royal Society two years later. Here, he met Isaac Newton, who was so impressed with the young man that, when the Chair of Mathematics at Edinburgh University became vacant in 1725, he offered to contribute towards MacLaurin's salary should he be appointed. Newton's offer was turned down, but MacLaurin was awarded this prestigious position.

Another figure of some importance was James Douglas, 14th earl of Morton. He came from a distinguished Scottish family, his father representing Lanark in Parliament in 1708, and the Orkney and Shetland Islands from 1713. James graduated from Kings College, Cambridge in

James Douglas, 14th Earl of Morton 1702–1768, and his family. A portrait by Jeremiah Davison. Image courtesy of the National Galleries of Scotland

1722, and succeeded to the title in 1738. His portrait, painted by Jeremiah Davison, can be seen in the National Portrait Gallery in Edinburgh. It was commissioned in 1740 and hung in the family's new home at Dalmahoy. In the painting, he stands modestly at the back, his arm resting beside a learned tome, allowing his family to fill the foreground. His son George, here depicted sitting on his mother's lap, died young, while his elder daughter, standing to the left, was dead before the portrait was completed. His wife Agatha died in 1748.

The earl of Morton was able to rise above these domestic tragedies with the help of his interest in science. He was elected to the Royal Society in 1733, observed the solar eclipse of 1734 with MacLaurin, and assisted the mathematician in establishing the Observatory in Edinburgh, not least with a contribution of £100 towards the project. He also raised money for James Cook's first voyage to Australia, and was rewarded by the sea captain who named Moreton Bay in Queensland after him, though he might have been a little disappointed with the spelling. The earl's love of science continued to the very end: he donated his body to the cause after his death, and it was anatomised by Sir John Pringle.

Such was Morton's standing that he became President of the Royal Society in London in 1764. He had been President of the Philosophical Society of Edinburgh from the time of its founding by MacLaurin in 1737. This Society was originally the Edinburgh Society for the Advancement of Medical Knowledge, but MacLaurin was keen to open it up to all the sciences. The Philosophical Society eventually became the Royal Society of Edinburgh.

Morton had inherited estates in Orkney, and travelled there in 1739, taking with him the talented optical designer James Short, one of MacLaurin's protégés, 'for the purpose of adjusting the geography of that remote archipelago' (Chambers, 1857). The trip was rather coloured by a serious dispute that arose between Morton and Sir James Steuart of Burray over feudal dues owed to the earl. The affair resulted in Steuart being fined and imprisoned. The matter rumbled on for years, and it may have been with some relief that Morton sold his interest in Orkney to Sir Laurence Dundas in 1766 for £63,000.

The pair returned to Edinburgh with reports that convinced MacLaurin of the need for a proper survey of these northern regions. His concern was not only with the inaccuracies of the coastlines of both the mainland and the islands, but also with the sea that separates them. The Pentland Firth is a notoriously dangerous stretch of water where the Atlantic meets the North Sea. With its strong tides, among the fastest in the world, leading to tidal races, currents, and eddies, it has over the centuries been the graveyard of many a ship that dared to pass through it. One whirlpool entered the sagas: called 'The Swelkie', it was said to have been caused by a sea-witch turning the mill wheels that grind the salt that is found in the sea.

MacLaurin's teaching at Edinburgh University had included work on surveying and cartography, which must have given him a pool of talented graduates on whom he could call when he needed work of that nature to be undertaken. For the project in the far north, he selected Murdoch Mackenzie to assess Orkney, and Alexander Bryce, the north coast.

Born in 1713, Bryce is another fine example of a Scottish Enlightenment figure. Described in the *Biographical Dictionary of Eminent Scotsmen* (1857) by Robert Chambers as an 'eminent geometrician', he graduated from Edinburgh in 1735. His mathematical skills led him into a number of areas. In astronomy, for example, he observed the transit of Venus in 1761 and 1769, and his *Account of a Comet observed by him in 1766* was published by the Royal Society. In 1776 he constructed what he termed an 'Epitome of the Solar System on a Large Scale' for the earl of Buchan at Kirkhill, one pillar of which can still be seen at Almondell Country Park, near Edinburgh (see colour section, page ii). For the Lord Privy Seal of Scotland, James Stuart-Mackenzie, he designed the observatory which occupies an impressive position on top of Kinpurney Hill in Perthshire. It housed amongst other devices 'an instrument contrived by him for ascertaining the magnifying power of telescopes, and

KINPURNEY TOWER, NEWTYLE   70409

The remains of the observatory on Kinpurnie Hill, designed by Bryce for James Stuart-Mackenzie. A photo by Valentine (c.1900).

a horizontal marble dial, made with great precision, to indicate the hour, the minute, and every 10 seconds' (Chambers, 1857). As far as surveying was concerned, he was involved in the construction of the proposed canal to join the Forth and the Clyde rivers, and the 'scales of Longitude and Latitude laid down agreeably to the observations of the Rev Bryce at Kirknewton Manse' were used by Andrew and Mostyn Armstrong in their 1773 *Map of the Three Lothians*.

His skills did not end there. In 1744 he was consulted by the magistrates of Stirling on the matter of supplying the town with water. He selected the site for the reservoir, and organised all the lead piping required for the project. As a reward, he was made a Freeman of the town. He dabbled in poetry, and in his younger days, wrote songs. He added three stanzas to the two already composed by Edward Mallet to the well-known song *The Birks of Invermay*, though the 1857 edition of Wood's *The Songs of Scotland* dismisses them as 'ludicrously artificial and nonsensical'. Bryce manages an astronomical allusion in the last verse:

Hark, how the Waters as they fall,
Loudly my love to gladness call,
The wanton waves sport in the beams
And fishes play throughout the streams.
The circling sun does now advance
And all the planets round him dance:
Let us as jovial be as they
Among the Birks of Invermay.

Nothing seems to have excited him more, however, than his discovery of the Standard Measure for weight and for liquid and dry measure in Scotland. On a visit to Stirling in 1750, he was told that the Standard Pint Jug was kept in the Council House. On studying it, he concluded that it was not the official measure. The magistrates, on being told, were 'ill able to appreciate their loss. It excited very different feelings in the mind of an antiquary and a mathematician' (Chambers, 1857). Bryce spent the next two years looking for it, without success, until the spring of 1752. Suspecting it might have been 'borrowed' by a brazier or coppersmith for making legal measures which they could then sell, his interest was aroused by a Mr Urquhart, a Jacobite rebel, who had failed to return after Culloden. Some of his belongings had already been sold, and the rest thrown into a garret, as it was thought to be useless. Upon looking there, 'to his great satisfaction, [he] discovered, under a mass of lumber, the precious object of his long research. Thus was recovered the only legal standard of weight and measure in Scotland' (ibid.).

Such, apparently, are the obsessions of a mathematical mind. In a state of triumph, he returned with the jug to Edinburgh, where, with the vessel placed exactly horizontal, with a minute silver wire the thickness of a hair laid across the mouth, and a plummet attached to each end, he poured water in until it touched the wire. It was then weighed. After 17 such trials, the Scottish Pint was proclaimed to weigh 54oz, 8 drops, and 20 grains, or, if you prefer it, 26,180 grains English Troy. Bryce continued to grapple with other measures, finding fault, for example, with the firlet, but let us leave him to his measuring, and simply acknowledge that 'for his good services to the City' (Chambers, 1857), he was made a Burgess and Guild Brother in 1754.

Whilst engaged in all these endeavours, Bryce was minister at Kirknewton, not far from Edinburgh. He had been ordained in 1745, and was presented to the parish by James, Earl of Morton, whose mansion was nearby at Dalmahoy. Bryce was known as a conscientious pastor, noted for his sermons. He dedicated his map to 'The Right Honorable the Earl of Morton, This map is inscribed by his Lordship's most obedient and most devoted humble servant'. Perhaps Morton was flattered by this, but the presentation to the benefice was certainly deserved, for the map was a magnificent achievement.

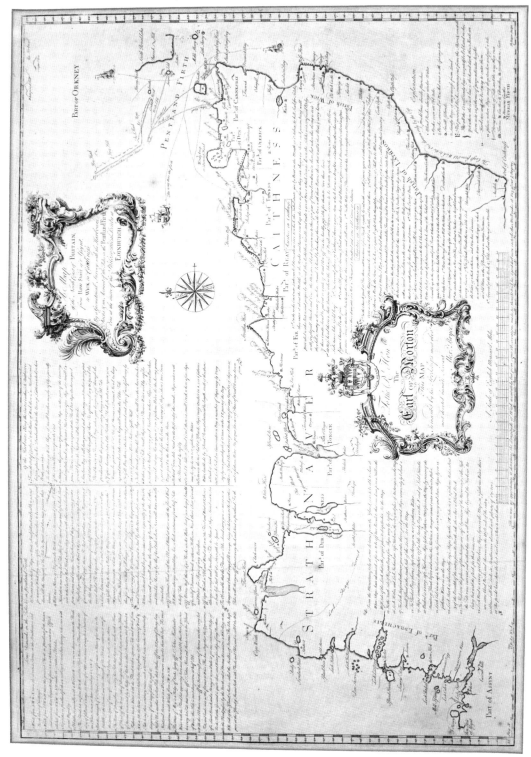

A Map of the North Coast of Britain from Row Stoir of Assynt to Wick in Caithness *by Alexander Bryce, 1744.*

Bryce was sent north by MacLaurin in 1740, ostensibly as tutor to a gentleman's son in Caithness, but the prime object was to draw up a map of the entire north coast. It was published in 1744 with the title *A Map of the North Coast of Britain, from Row Stoir of Assynt to Wick in Caithness, By a Geometrical Survey with the Harbours, Rocks, and an Account of the Tides of the Pentland Firth, done at the desire of the Philosophical Society of Edinburgh.*

There are sadly few details of exactly how the survey was carried out. It took three years to complete. Bryce was said to have been accompanied by three assistants, often under threat 'from the peasantry (which made it necessary for him to go always armed) who did not relish so accurate an examination of their coast, from motives of disloyalty, or because they were afraid it would deprive them of two principal sources of income – smuggling and plunder from the shipwrecks' (Chambers, 1857). The scholar Diana Webster believes Murdoch Mackenzie was at times employed on the survey, which involved triangulation techniques. Such experience would have stood him in good stead later for his own superb work in the vicinity, which resulted in the charts of the *Orcades* published in 1750.

Bryce's map is a large sheet, with the engraved area measuring 69cms x 48cms; much of it is filled with annotations as well as the outline of the coast at a scale of 1 inch to 3 ½ miles. Ben Hope and Ben Loyal are named, but otherwise little detail is shown inland: only the occasional mountain profile that would be visible from the sea, for this is a coastal chart. The emphasis

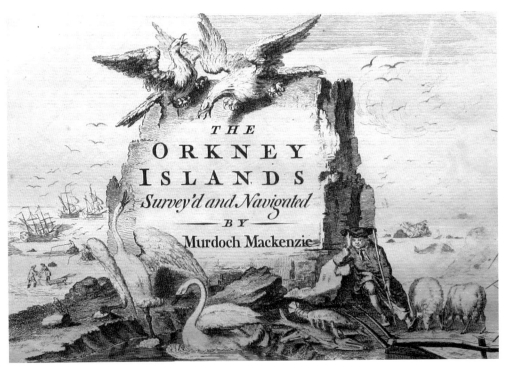

The splendid title cartouche to Mackenzie's Orcades, *a set of eight maps published in 1750.*

is on navigation and sea travel, with 'the way thro the Firth' being plotted south of Orkney. The fearsome tides are also marked. For example, 'The Boar of Dungsby is a very hot Tide that... shoots over to the island of Stroma'. Obstacles are pointed out, such as the ridge of rocks known as the 'Men of Mey' which lie 'under water with the flood, but appearing here and there at Ebb'. Small anchor symbols show 'Harbours and the best way to enter them', and at Scrabster he notes 'for the better Securing of Ships, there's two iron rings fixed to the rocks on the shore'. John Knox spotted these same two rings on his tour 40 years later, commenting in alarm and astonishment that they were 'the only aid to shipping between Ireland and the Baltic' (Knox, 1785). Bryce also found room on the map for a list of 'Rarities' both in Caithness and Sutherland. He notes the carved stone which still stands in Farr churchyard 'rais'd', he claims, 'over a son of the King of Denmark, who was killed in Battle and buried there'. He also includes Dun Dornadilla south of Eriboll, where 'between the outer and inner walls are Beds for the Hunters and their Dogs, with places for Bows and Arrows', and Smoo Cave, near Durness, which had an inner loch that had not yet been fully explored, though 'some have had the Courage to try it with Boats and Candles.'

The quality of Bryce's map was such that, more than 60 years later, the cartographer Aaron Arrowsmith pronounced it to be 'very accurate, after a minute examination ...' (Chambers, 1857). The greatest achievement is the coastline itself, at last plotted

Contemporary mapmakers were quick to incorporate Bryce's work. This is the title cartouche of a map by Emanuel Bowen issued in 1747. It fully acknowledges the efforts of various surveyors including Bryce and Mackenzie, saying of the latter's Orcades 'A very useful design highly worthy of encouragement.' Bowen was the engraver of Mackenzie's maps when they were issued in 1750.

"Cape Wrath". A particularly dramatic, if not exactly accurate impression by W.H. Bartlett, from Beattie's Scotland, *1838.*

accurately. He admits that 'The Coast from Old Wick to the Ord [was] drawn only by the Eye', suggesting the rest was done geometrically. All the little settlements along the coast are located, the coastal indentations are complete, and the little islands off the west coast, together with those above Loch Eriboll and the Kyle of Tongue are in their rightful places. Above all, there is at last the substantial block of land to the west of Faraid Head and the Kyle of Durness, terminating at Cape Wrath. What a block of land it is, too, as described by John MacCulloch:

> As if nothing else could resist the fury of a northern ocean, Nature seemed to have reared a huge and rude barrier which neither storms nor waves should ever have the power to move. I felt how insignificant Cape Rath would have appeared, how Nature herself would have erred, had Britain here terminated in any other manner, had any lower or tamer point of land been opposed to this raging sea. Here she was truly her own poet; nor could the most vivid imagination and the most correct taste, have conceived a more thoroughly harmonious adaptation of character: that of the wildest land to the wildest ocean, the strong-built and immovable rocks to the furious waves; to the majestic breaking of the lofty billows,

a still more majestic pyramid, towering far above their greatest efforts, and, as the termination of the rude mountain ranges of Scotland, a buttress worthy of all their grandeur and all their strength (II, 1824).

It had taken exactly 180 years to have this magnificent buttress acknowledged on a map.

Bryce returned to Edinburgh in 1743. His knowledge of the geography of Scotland enabled him to give advice to Cumberland's Hanoverian army as it headed north in 1746. MacLaurin was also a confirmed anti-Jacobite; he helped to prepare the defences of Edinburgh as Bonnie Prince Charlie advanced on the city in 1745. The Jacobite victory at Prestonpans meant that MacLaurin had to make a hasty escape south, where he took refuge in York. 'Here I live as happily as a man can do who is ignorant of the state of his family, and who sees the ruin of his country' (letter to a friend, 1746, quoted in *The Georgian Era*, 1834).

At MacLaurin's request, Bryce took over his teaching when in 1746 he became too ill to carry out his duties, and he died soon after. Bryce, the poet, was moved to compose this eulogy in his memory:

> You angel guards that wait his soul,
> Amaz'd at aught from earth so bright,
> Find nothing new from Pole to Pole,
> To show him in a clearer light.
>
> Joyful he bears glad news on high
> And tells them through celestial space;
> See Newton hastens down the sky
> To meet him with a warm embrace.
>
> The list'ning choirs around them throng,
> Their love and wonder fond to show,
> On golden harps they tune the song
> Of nature's laws in worlds below.
>
> O Forbes, Foulks, loved Morton, mourn;
> Edina, London, Paris sigh;
> With tears bedew his costly urn
> And pray – Earth light upon him lie.
>
> (Chambers, 1857)

CHAPTER TWO

A NEW SURVEY OF THE WHOLE

The Highlands are but little known even to the inhabitants of the Low Country of Scotland, for they have ever dreaded the difficulties of travelling among the mountains; and when some extraordinary occasion has obliged anyone of them to such a progress, he has generally speaking made his testament before he set out, as though he were entering upon a long and dangerous sea voyage, wherein it was very doubtful if he should ever return.

(Burt, 1818)

Culloden. 16 April 1746. A dismal, dank morning, as befitted the scene of the last pitched battle on British soil. The rebel Jacobite Army, under Charles Edward Stuart, or Bonnie Prince Charlie as he was popularly known, was no longer the tight, purposeful body that had defeated the English at Prestonpans, and caused such consternation in London as it advanced as far south as Derby. Now, after weeks of retreat northwards, pursued by the Hanoverian Army under the Duke of Cumberland, third son of George II, the rebels were cold and hungry, and the organisation in chaos, with men deserting back to their Highland homesteads. Even as they waited in formation on the evening of the fifteenth, their promised meal failed to materialise. It was stuck in Inverness, with no transport available to bring it to the waiting troops.

The choice of battleground was a poor one, 'not proper for Highlanders' (Lord George Murray, from his journals, quoted in Prebble, 1967). The fearsome charge of the Jacobites, which had proved so successful at Prestonpans, was repulsed at Culloden by the muskets of the disciplined Hanoverian ranks. In less than one hour, the insurgent forces were routed and in chaotic retreat, leaving 2,000 of their compatriots dead or wounded on the battlefield.

The consequences of the battle were enormous. The survivors were hounded, imprisoned, hanged, or transported at the whim of their conquerors. The Highland population as a whole, whether innocent or guilty, suffered the indignity and pressure of the Hanoverian

"The End of the 'Forty-Five' Rebellion". An engraving by W.B. Hole.

army in its midst. But, above all, the very structure of Highland society was threatened. The victory had been so comprehensive that the victors, in a state of fear and triumph, sought to destroy the Highland way of life.

'To the southern inhabitants of Scotland, the state of the mountains and the islands is equally unknown with that of Borneo or Sumatra. Of both they have only heard a little, and guess the rest'. So wrote Dr Johnson on his journey to the Western Isles in 1773, sentiments that were echoed by the Welsh antiquarian and naturalist Thomas Pennant who thought the Highlands 'almost as little known to its Southern brethren as Kamtschatka [*sic*]' (1793). The mountains which rise to the north of Stirling and Perth formed a natural barrier beyond which, because of its remoteness, was to be found a very different way of life. The inhabitants even spoke in a strange tongue, Gaelic. Keats, on his trip to the region in 1818, found himself 'for the first time in a country where a foreign language is spoken' (letter to Tom Keats, 20 July; Keats, 1901).

Since 1707 and the Act of Union, Scotland had been further united with England, the two countries now 'One Kingdom under the Name of Great Britain'. To those in the Highland regions, the king of England was an irrelevance. They may not even have known of the Union. Their society had, for centuries, been run on traditional clan lines, and their allegiance lay with their local chief, the laird. Writing at the end of the nineteenth century, Osgood Mackenzie remembered the innkeeper of Kinlochewe who 'would cheerfully have gone to the gallows if she were quite sure that would please the laird' (1949). Other reformers like John Knox saw the situation less romantically, likening the chiefs to 'West India Planters' who 'command the labour of their slaves' (Knox, 1787). The Gaelic word *clann* means 'children', and the relationship with the laird was paternalistic. He would let out portions of land to his tenants, and look after them at times of hardship. In return, they would give him a percentage of their produce (if there was any), and some of their time (often the best weeks of the year) for harvesting, building, or any chore he demanded. At times of conflict, he could also call on them to take up arms.

Had Charles Stuart been able to call on the support of all the Highland clans when he raised his father's standard at Glenfinnan in 1744, he would have had at his command an army of 30,000 – rather more than the 6,000 who managed to reach as far south as Derby and threaten London. However, the Jacobite rebellions of 1715, 1719, and 1745 did not simply set Scotland against England, nor even Highlanders against Lowlanders. Some Highland Chiefs were staunch supporters of the Hanoverian cause. One such was Lord Reay of the Sutherland Mackays.

The Young Pretender looked to France for support. In March 1746, a ship named Prince Charles left Dunkirk for Scotland. Originally called HMS Hazard, a 14-gun Merlin-class

"Passage of the Highland Army along the Side of Loch Eil, 1745". An engraving by T. Allom.

TONGUE HOUSE.

Tongue House, from an early 20th century photograph.

sloop launched in 1744, it had been captured by Jacobite forces at Montrose and renamed. Now, as it headed for the rebel army, it was laden with £13,000 in gold, arms, and other supplies. The money was much needed, not least to pay the troops who by this time were showing signs of restlessness and dissatisfaction. As the ship approached the coast, it was spotted and chased by a British naval vessel, the Sheerness. Rounding the north coast, shots were fired, and the Hazard was forced into one of the big inlets, the Kyle of Tongue. This was an unlucky choice on two scores: the Kyle of Tongue contains pronounced sandbanks, and the village of Tongue was the local Hanoverian headquarters. Lord Reay watched from his house on the east shore of the Kyle as the sloop ran onto the shoals and stuck fast. The French and Spanish crew abandoned the vessel, taking the gold with them, and headed south down the Kyle for what they perceived to be the safety of the mountains. Before they could reach them, they were intercepted by a force of 80 men from Tongue. Five of the crew were killed, and the rest taken prisoner. Not all the treasure was retrieved, however. At least one chest went missing. Local rumour has it that it was flung into Lochan Hakel, which lies beneath Ben Loyal. From time to time, a cow emerges from the water with a golden louis d'or in one of its hooves, but the hoard has yet to be found.

Had the fleeing crew made it to the mountains, they would have faced a formidable journey before meeting up with the rebel forces. The nearest road was 100 miles to the

Loch Hakel, with Ben Loyal in winter array behind.

south, and between lay a vast area of almost uninhabited heath and boggy moorland. To traverse the Highlands at this time required local knowledge and a guide. No map published in the first half of the eighteenth century was of any use as a navigational aid. Only the local inhabitants knew what was there, and nobody was more aware of this than the British Military Command.

The relief felt in England at the news of the victory at Culloden was considerable. Huge crowds greeted Cumberland as he rode down Tyburn Lane on 25 July, seven days after leaving Fort Augustus. 'Sweet William' he may have been to his admirers, but he quickly acquired the nickname 'Butcher Cumberland' once the ferocity shown towards the defeated in the aftermath of the battle became apparent. With the victory came a determination to eliminate anything that made the Highlands so separate and alien a part of the British Isles.

Attempts to achieve this had already been made after the failure of the first Jacobite rising in 1715, when Charles Edward's father endeavoured to regain the throne lost by his father, James II of England. A Disarming Act had been passed, which forbade anyone in specified parts of Scotland to carry a 'broad-sword or target, poignard [long knife], whinger [type of dagger], or durk [short dagger], side pistol, gun, or other warlike weapon' unless authorised. This proved difficult to enforce, as is clear from the arms available to the rebel army of 1745. Following Culloden, an Act of Proscription was passed on 1 August 1746, which went a good deal further: not only were weapons banned (with harsher penalties, including transportation, imposed for those caught armed), but also all forms of Highland dress. The list in the Act includes 'the Plaid, Philabeg or little kilt, Trowse, Shoulder-belts, or any part whatever of what particularly belongs to the Highland Garb', and specifies that 'no tartan, or partly-coloured plaid or stuff shall be used for Great Coats or upper coats…'. It is not known how well these restrictions were enforced. The *Tain and Balgowan Documents* record that in 1751, 'Wm. Ross, son of Alexr. Ross in Dalnacleragh, now prisoner in the Tolbooth of Tain, has been taken up and incarcerated for wearing and using the Highland dress and arms' (Macgill, 1909), while Captain John Forbes, in his Report on the Annexed Estates (those confiscated by the Crown following Culloden) states that the factor at Coigach had 'apprehended a fellow in a tartan coat, whom he sent to Dingwall jail, where he remained till liberate in the course of law'. But the general feeling was that in the more remote areas 'some fellows may presume to transgress where they can do it with impunity' (Wills, 1973). The Reverend James Hall, travelling up the east coast in 1807, observed, 'The Highlanders, being compelled to lay aside the kilt, or philibeg, after 1745, and to have each man a pair of breeches, provided themselves with breeches, though they did not always wear them, but used the old kilt at home, and when they went abroad, carried their breeches, swung over a stick, resting on their shoulders' (1807). Other ploys included sewing their kilts up the middle so that they became culottes. There are plenty of portraits and paintings dating from the years of Proscription which suggest that the traditional garb was still quite publicly displayed – the aristocracy dressed proudly in the most elaborate tartans, while humbler folk are portrayed dancing to the bagpipe and fiddle. However, the Frenchman Alexandre de la Rochefoucauld, travelling through the region in 1786, was struck by the effect the Act had had on the population. He thought that, with the removal of their traditional, distinctive clothing, 'they faced becoming a former people' (Scarfe, 2001).

Further restrictions followed the Act of Proscription. Negative attitudes to Gaelic predated the Jacobite rebellion, which exacerbated opposition to the language; for a short time afterwards, Gaelic was suppressed in schools. Even more devastating was the Heritable Jurisdictions Act of 1746, which abolished the judicial powers of the clan chiefs that had regulated Highland society for so long. The effect of this was to be felt increasingly over the next 100 years. Thomas Telford put it succinctly when he observed, 'The Lairds have transferred their affections from the people to flocks of Sheep' (Rolt, 1958).

Once these Acts had been passed, the onus fell on the military to enforce them and to maintain the peace. In the immediate aftermath of Culloden, soldiers found themselves traipsing through soggy, rocky, mountainous terrain, without having much idea of where they were, looking for survivors of the battle, and in particular for the Young Pretender himself. The unpleasant General Henry Hawley, when sent north to take command of government forces, complained, 'I am going in the dark. Marechal Wade won't let me have his map' (Letter, 1794, quoted in Moir, 1973). What exactly this map was is unclear – it appears to have been a manuscript chart, for he wished it 'was either copied or printed'. The most detailed map at that time was by the military engineer John Elphinstone. It was *A New and Correct Map…laid down from the Latest Surveys,* and he acknowledged various sources including John Adair, Alexander Bryce, and Murdoch Mackenzie. But he admitted that his map still had faults, though it was 'as Correct as possible till a New Survey of the Whole [could be] made'. Much of the west coast was far from accurately plotted, and, in any event, at a scale of 13 miles to the inch, Elphinstone's map was never going to be of much use as a detailed navigational aid.

Another man who knew the need for 'a new survey of the whole' was David Watson, who had been appointed Deputy Quartermaster-General in North Britain by the Duke of Cumberland shortly after Culloden. His main task was to look after the army's provisions, and he found himself forced to petition his commander, explaining that he and his colleagues frequently 'found themselves greatly embarrassed for want of a proper Survey of the Country' (John Watson, *Memorial,* quoted in Hewitt, 2010). Cumberland took it upon himself to approach George II in person on the subject, and as a result it was resolved to make a 'Compleat and accurate Survey of Scotland' (ibid.).

Unfortunately, the official records as to how things proceeded from this point are lost. We do not know, for example, why Watson chose an unknown 21-year-old called William Roy to oversee this ambitious survey. There is some evidence to show that Watson knew Roy's father – he was a factor, a job requiring expertise in surveying. One imagines Roy junior was taught this skill from an early age by his father, and Watson may have taken note of his precocious ability. Roy spoke affectionately of Watson as a man 'active and indefatigable, a zealous promoter of every useful undertaking, and the warm and steady friend of the industrious' (*Account of the Measurement of a Base on Hounslow Heath,* 1785, quoted in Moir, 1973). Watson put Roy's industry to the test in the vicinity of Fort Augustus, and after two years of work, the 'specimens of his progress were so satisfactory that it was determined to extend the Survey over the whole of the north of Scotland.' Thus wrote Aaron Arrowsmith, in a memoir published in 1809 to accompany his own improved *Map of Scotland.* His research had led him to talk to some of the soldiers who had worked on the survey, and with the official records lost, this memoir is the best source available.

Roy's appointment was an inspired choice. He went on to rise through the ranks to the position of general, and was a crucial figure in the formation of the Ordnance Survey at the end of the eighteenth century. In fact, four other members of the team also became generals (Watson, Hugh Debbeig, Charles Dundas, and Charles Tarrant). It was clearly a talented group, complemented by the appointment of the 17-year-old Paul Sandby, who went on to become an eminent artist in his own right, as Chief Draughtsman. His influence can be seen on the design of the 'Great Map', as it became known, but of equal importance are the images he produced of the surveyors at work in areas of the Highlands (at Kinloch Rannoch, for example) that had rarely been visited before by artists. There are also sketchbooks by Sandby in which scenes of ordinary people at work, of soldiers on duty, and even one of the office of the survey with the Board of Ordnance assembled are depicted. These form a valuable contribution to the archive of images of life in the Highlands at this time.

Surveying team at work in the Highlands.
An engraving by Paul Sandby, 1765.

The survey was carried out by six teams, each consisting of eight men led by a surveyor. They worked during the summer months, spending the winter in the Ordnance Drawing Room in Edinburgh, which was probably in the Governor's House in the castle. Sandby was required to draw up presentation maps from various sketches, while on the actual general survey, he was, according to Arrowsmith, responsible for the 'Mountains and Ground' (*Memoir Relative to the Construction of the Map of Scotland*, 1807, quoted in Moir, 1973). The use of hatching as employed here (shading in fine lines) was to become the standard way of depicting high ground on Ordnance Survey maps until replaced by contour lines in the early twentieth century. Some mountains, such as Suilven, are

Detail from General Roy's "Great Map" showing a bird's-eye view of the mountain Suilven in Sutherland. Image courtesy of the British Library.

depicted in birds-eye view. Certainly, the 'Great Map' is visually one of the most remarkable ever produced. Like Pont's survey of 150 years earlier, it did not aim to provide complete topographical accuracy: it was rather what Roy called 'a magnificent military sketch than a very accurate map of the country…No geometrical exactness is to be expected, the sole object being, to shew remarkable things, or such as constitute the great outlines of the Country' (Roy, 1785). The map was never published in its day, but it is now available in atlas form (https:// maps.nls.uk/geo/roy and Roy, 2007). Open the book anywhere and you will see cartographic images like no others: as much a work of art as a map, the north-west appears like some moonscape, an abstract painting with the high ground represented by random blobs on the page. No other map depicts so clearly the vast isolation of this northern expanse, with tiny hamlets occasionally clinging to some fertile glen in the midst of a sea of mountains.

So successful was the survey that it was carried on in the south of Scotland, though a fair copy of this section was never made. The work was opportunely completed in 1755. The war in America, which involved Britain and France, produced a fear of invasion from across the Channel, and the surveyors in Scotland were called to the south coast for detailed work in Kent and Sussex.

With the lack of any first-hand accounts and descriptions, the work involved in making this survey can only be imagined. The vagaries of the weather need no emphasising. The equipment – the large theodolite for surveying angles, the chains for measuring distances, all the tents and baggage for outdoor living – had to be lugged around, with much of the work being carried out from mountain tops. Moreover, the sight of soldiers in groups at work would, in the years following Culloden, have been viewed with much suspicion in some areas, and like Bryce, they must have faced local abuse and opposition.

The work on the 'Great Map' may have been terminated in 1755 when the surveyors were summoned south, but Roy's interest in the mapping of North Britain did not end there. He had allowed his fascination with history, archaeology and the Roman occupation to go hand-in-hand with the more general surveying work, and, particularly when in southern Scotland, had made sketches of the various sites of antiquities that he came across. In 1764 he discovered a new site at Cleghorn, and over the next few years worked on numerous plans for what became the *Military Antiquities of the Romans in North Britain*. His superb map of Scotland found in this volume incorporated measurements and details made during the military survey.

Roy never actually saw the volume in print. It was published by his colleagues and friends in the Society of Antiquaries in 1793, three years after his sudden death. Like Pont, Roy remains a shadowy figure. Although he was a man who rose through the ranks to become a Major-General, who oversaw the first accurate triangulation survey from London to the Kent coast, and who is credited with being a major force in the foundation of the Ordnance Survey, there is no known portrait of him. There is a caricature drawing of *The Antiquarian*

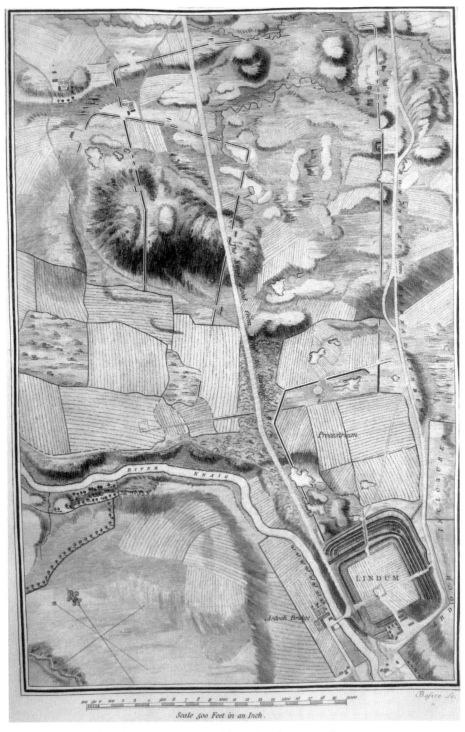

Plate X from Roy's Military Antiquities. *The style of mapping has much in common with that found in the Great Map of Scotland.*

Society by George Cruikshank (see colour section, page iii), in which is depicted amongst the various members a figure on the right, his pockets stuffed with papers marked 'Ordnance Affairs'. He is eyeing a 'Roman Vase' on the table in front of him. If this is William Roy, it is the best likeness we have, though it was drawn some 20 years after his death.

The accurate mapping of Scotland continued at a slow pace during the second half of the eighteenth century. The landmark maps (those that include some genuine, new survey work rather than being mere copies of someone else's research) include one by James Dorret which was published in 1750, and consisted of a large plan on four sheets, with two side-strips. This was hailed as a great improvement and remained the standard blueprint until those issued by Cary and Arrowsmith in the 1790s. With road improvements becoming increasingly common further south, hand-in-hand with an increase in coach travel generally, cartographers began to add roads as a matter of course. In the far north, Dorret depicts a clear road heading up from Dornoch, cutting up through the centre of Sutherland all the way to Tongue, where it meets a road coming from Thurso along the coast, continuing

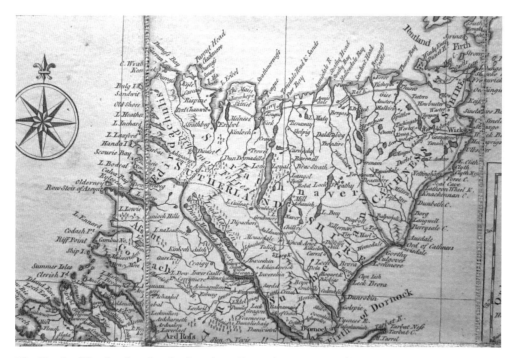

The North of Scotland, a detail taken from Dorret's map, 1750. (An edition by Robert Sayer, c.1765). This was hailed as a great improvement on previous maps, but it depicts roads that simply did not exist.

westwards over the Moine, a notoriously boggy tract of ground, round Loch Eriboll and on to Durness. Alas! These roads simply did not exist, and would not do so for another 70 years, in some cases longer. Roy's map does show a track heading south from Tongue, but it peters out at Loch Loyal. There is also a track along the north coast to Thurso, but as late as 1820, the edition of Duncan's *Itinerary* warns, 'There is no made road further than two miles beyond Reay Kirk. Travellers going in the direction of Tongue and Durness should endeavour to procure a guide, as without one he may deviate from the path, which is only what the Highlanders call a *bridle road*, and few houses to be met with. The traveller will do well to fill his *flask* and supply his *scrip* [a wallet or bag] at Reay Kirk Inn, as he may rest assured he will require their aid before he reaches Tongue.' It adds encouragingly that 'the natives he will find very hospitable.'

Mapmakers were not slow to criticise each other. Playfair, in his 1819 *Geographical and Statistical Description of Scotland* includes a map which has its own faults (such as roads that did not exist), but he does not hesitate to belittle the efforts of earlier cartographers. The examples by 'Adair … Sanson, and Elphinstone, being inaccurately constructed and wretchedly engraved, are now consigned to oblivion.' He goes on to complain that Dorret's map 'is inaccurate, and has not been materially improved by Kitchen and Bowes. Armstrong's *Scots Atlas*, though neatly engraved, is little valued. Instead of an accurate survey which he pretends to have made, he has copied the errors of others, and ingrafted mistakes of his own'. However, he admires the work of Murdoch Mackenzie, Bryce, and Arrowsmith, commenting with regard to the latter that 'the inaccuracies and defects of this map, which are few, will be corrected and supplied by the trigonometrical survey that is now almost completed.'

This trigonometrical survey was another military affair. Realistically, detailed mapping of rugged and roadless terrain like that found in the Highlands was always going to rely on the Ordnance department, which could provide the workforce required. This particular project was not the full Ordnance Survey; that was to come later (much later in the case of the northern areas). Rather it was part of a project to establish accurately the measurements of an entire meridian arc through Britain – in effect, a long baseline from which other trigonometrical measurements could be made. This was all part of the demand for increased accuracy that was a feature of late Georgian cartography.

The man chosen to lead this survey, Captain, (later General), Thomas Colby, was certainly not the type to be put off by small details such as high mountains and a lack of roads. The tone was set even as he travelled up to Scotland. On the stage coach from London 'neither rain nor snow, nor any degree of severity in the weather, would induce him to take an inside seat, or to tie a shawl round his throat; but, muffled in a thick box coat, and with his servant Fraser, an old artilleryman, by his side, he would pursue his journey for days and nights together, with but little refreshment, and that of the plainest kind,– commonly only meat and bread, with tea or a glass of beer' (Portlock, 1869). Osgood Mackenzie's uncle described him as 'a highly educated man of science, from astronomy all the way downwards, full of every kind of information, and most able and glad to pass

"Glencoe from the SW". A cartoon from The Graphic, *by W. Ralston, 1875.*

it on to others' (Mackenzie, 1949). It is, however, his physical attributes that strike one most. His biographer, Joseph Portlock, describes him as 'possessed of a singularly nervous and elastic frame, which no fatigue could overcome. Exposed almost bare-breasted to the storm, he appeared unaffected by the bitter blasts of winter, and day after day he persevered in walks over mountain districts which no ordinary strength could have mastered'. Up in the Highlands he thrived, probably to the dismay of his men at times, on 'that very wildness, from that glorious sense of freedom which seems to swell the bosom as the fresh mountain air is inhaled, a charm which dispelled from the performance of duty the very thought of labour'.

He was not a perfect physical specimen: in 1803 a pistol he was holding exploded, shattering his left hand, whilst one of the fragments pierced his skull. The accident might well have killed a weaker man, but he survived, though without his left hand, and with a permanent scar on his head.

He was clearly a man to lead from the front. His biographer speaks of his 'utmost straight-forwardness in every proceeding of the survey', and one of his surveyors, Robert Kearsley Dawson, states that even in the difficult and exhausting business of getting all their equipment up the mountains, 'Captain Colby invariably took an active part; and he would never sit down to rest or take any food or refreshment until everything had been satisfactorily accomplished and secured'. Dawson's father had been a well-known surveyor in his own right, and had been with Colby when the pistol exploded. It was he who saved Colby's life by immediately obtaining medical aid. For this we owe him a debt of gratitude, as we do to Dawson junior, for he provided an account of his time with Colby on the survey, which appears in full in the biography. It gives us a detailed and unique insight into the life of an Ordnance surveyor.

Dawson remembers that, when perched up high on a mountain, 'it was no uncommon occurrence for the camp to be enveloped in clouds for several weeks together, without affording even a glimpse of the sun or the clear sky during the whole period. And then in a moment the clouds would break away or subside into the valleys, leaving the tips of the mountains clear and bright above an ocean of mist, and the atmosphere calm and steady, so as to admit of the observations for which the party had waited days and weeks to be taken in a few hours.' Woe betide any local gentry or interested caller who might visit them at one of these favourable moments. Whilst in periods of inactivity, Colby could be charming and welcoming, only too keen to show off all the instruments that were ready for action, yet 'nothing appeared to worry him more than the approach of visitors when we were really at work, and whatever might be their rank, he would then scarcely speak to them or show them even common attention.'

The weather and conditions could be most capricious. On midsummer day 1819, after a stormy spell in the Cairngorms, the visibility was so good that Colby was able to pick out through the telescope a brig under sail that he calculated to be 100 miles away, to the north. One week later, they suffered a hailstorm, followed by snow which fell for at least three hours. The men took to snow-balling to keep warm.

The most remarkable feature of Dawson's account is that it reveals the huge distances they covered on foot, and the speed at which they worked. He described one expedition in detail – a journey in search of observation points that might be of use to the Ordnance Survey in the future. They were to explore the country to the west and north-west of their camp on

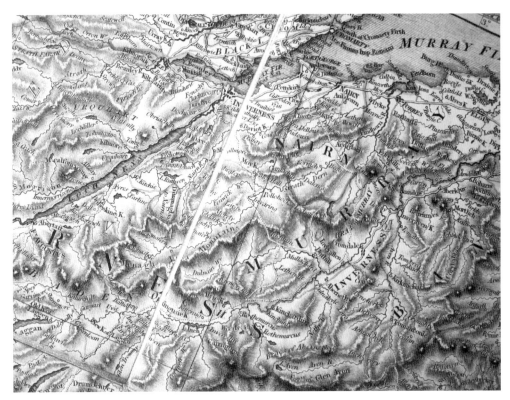

Detail taken from a map of Scotland by William Faden published in 1813. It shows the area traversed by Colby and his party on the first two days of their trek to Skye. Corriehabbie is marked near the lower right hand edge of the map. They proceeded in a westerly direction towards Cromdale (Grantown is not marked), thence southwest to Aviemore (the Inn is marked). The next day, the toughest for Dawson, they headed into the Monagh Lea Mountains, rather than take the road down to Kingussie. Carn Dearg is one of the hills to the north of Garviemore Inn, shown on the bottom left of this map. It was here that they spent two nights, climbing above Loch Laggan on their 'day off'. Interestingly, the map does not show Wade's road over the Corrieyairack pass down to Fort Augustus.

Corrie Habbie (Corryhabbie) on the northeast margins of the Cairngorms. Colby had in fact just completed one such investigation, returning to his base on 21 July having walked 513 miles in 22 days. Two days later, on 23 July, he was off again with a fresh party of soldiers that included Dawson. The first port of call was to be Grant Town (Grantown-on-Spey), a distance of 24 miles: 'Captain Colby having, according to his usual practice, ascertained the general direction by means of a pocket compass and map, the whole party set off, as on a steeple-chase, running down the mountain side at full speed, over Cromdale, a mountain about the same height as Corrie Habbie, crossing several beautiful glens, wading the streams which flowed through them, and regardless of all difficulties that were not absolutely unsurmountable on foot. Sometimes a beaten road would fall in our course, offering the temptation of its superior facilities to the exhausted energies of the weary members of our party; and in such cases freedom of choice was always allowed them. Captain Colby would even encourage such a division of his party and the spirit of rivalry which it induced, and took pleasure in the result of the race which ensued'.

They arrived at Grantown-on-Spey five and a half hours later, which means they had averaged over four miles per hour, a remarkable speed given the terrain. They dined here, but their day was not yet over; they followed the high road to Aviemore, another 15 miles further on, which made a total for the day of 39 miles.

The next day, Dawson was struggling from the moment he woke up. 'Started at nine o'clock. I was dreadfully stiff and tired from the previous day's scramble, and with difficulty reached Pitmain (13 miles) to dinner'. The rest of the day was to be spent in getting to Garviemore Inn (Garvamore), a distance of 18 miles by road; poor Dawson was not sure that he could manage it. But worse news was to follow. Colby had no intention of using the road, but rather they would head cross-country for Cairn Derig (Carn Dearg? 3,087 ft), which he calculated to be 3,500 feet high and ten miles distant. From there, having built a cairn on the summit, they would complete the journey to Garviemore. This would add considerably to the 18 miles by road, the entire distance over the familiar rough, boggy terrain. Desperately, Dawson begged to be allowed to join two colleagues who had been given permission to use the road on account of their blistered feet. 'Captain Colby would not excuse me, and I had no alternative but to make the attempt, feeling sure that I should eventually be left upon the ground or carried home upon the men's shoulders.' However, to his relief he found that Colby was right: he kept pace with the group, and by the end, found himself in remarkably good condition. 'The second day on such a journey is generally the worst, but the first had broken me in.' What is more, he claimed that he 'never experienced anything like fatigue throughout the remainder of the excursion.' This was just as well as the expedition maintained this pace for the remaining three weeks.

The next day, Sunday 25 July, was a day of rest – what better way to spend it than climbing Bui-Annoch, 4,020 feet above Loch Laggan, where they enjoyed as good a view as any in Scotland? It seems as though Dawson had a tendency to overestimate the heights of these mountains. Bui-Annoch is not marked on modern maps. It is almost certainly the ridge now marked as A' Bhuidheanach, which leads to the summit at Creag Meagaidh (3,700 feet). Perhaps he overestimated distances too, but he states that they had covered 24 miles on their day off. On Monday they were back up to 40 miles, this time by road over the famous Corrieharrack (Corrieyairack) pass down to Fort Augustus. Dawson begins to express various comments and sentiments that we will come across repeatedly with other travellers to these remote Highland regions. 'A dreary and melancholy journey' he calls it, passing houses deserted after the clearances, and ending up at a poor inn at Cluny, where his bed consisted of four chairs 'placed as evenly as the earthen floor would permit.' They were served a supper of salmon that was so stale that they were unable to eat it. They did, however, find solace in that great Highland stand-by, 'a mess of oatmeal-porridge with goat's milk'. In contrast, breakfast the next morning at Invershiel found salmon 'almost alive from the fishery in front of the inn.'

They then crossed over to Skye, where they attempted to climb the Cuillin on 29 July. Having no ropes, they failed to ascend the Inaccessible Pinnacle, that final obstacle that has alarmed many a Munro-bagger. They spent several hours trying to find a way up, but to no avail – 'the only instance in which Captain Colby was ever so foiled.' However, they delighted in the perpendicular ridges, on one occasion shuffling along seated, with vertiginous drops on either side, and returned to the inn 'gratified above measure with what [they] had seen, though disconcerted with [their] professional failure.' On 31 July they climbed Scour-na-Marich (Sgurr a' Mhaoraich?) before breakfast, and for the rest of the week enjoyed intense heat which, of course, was accompanied by 'baneful attacks' from midges. 'Our arms, necks, and faces, were covered with scarlet pimples, and we lost several hours' rest at night from the intense itching and pain which they caused.' Another, less satisfactory, adventure involved trying to resuscitate 'a poor shoe-maker, who, after a long day's work in repairing our men's shoes, had gone to bathe and got out of his depth.' Captain Colby, of course, 'at once undertook the management of the case.' Aided by what Dawson calls 'the ordinary rules of the Humane Society', his administration of first aid was to no avail, and the man died.

They enjoyed the hospitality of the Mackenzies of Gairloch, from where they explored Loch Maree, then on to Beauly, and Inverness, arriving back at Corrie Habbie on Saturday 14 August. They had covered '586 miles in twenty-two days, including Sundays, and the days on which [they] were unable to proceed from bad weather.' They deservedly enjoyed a supper that night consisting of boiled salmon and grouse, with 'vegetables and fruits of various kinds of the finest quality; ale, porter, and wine', all provided by the Duke of Gordon, who was then at Glen Fiddich.

These treks were often undertaken in July and August because these months were notoriously bad for 'seeing' – the warmer weather caused more atmospheric disturbances. September was the best month for surveying, and the expedition made the most of this, finishing their observations on Wednesday 29 September 1819. Blessed with continuing fine weather, they were able to take the tents down and pack them up dry, while the great theodolite, a huge piece of equipment which had been made for the Ordnance Survey by Jesse Ramsden, was loaded onto its wagon and sent south by road. As was the custom, a farewell feast was proposed. 'The chief dish on such occasions was an enormous plum-pudding. The approved proportions of the ingredients being, as we were told, a pound of raisins, a pound of currants, a pound of suet, &c. to each pound of flour; those quantities were all multiplied by the number of mouths in the camp, and the result was a pudding of nearly a hundred pounds weight.' This was suspended over a fire, and kept boiling for 24 hours, guard-duty being organised to maintain the fire and ensure a constant supply of water. Whether this delicacy was enough to replace the calories the party had used up over the past few months, Dawson does not say, but they raised their glasses with enthusiasm, the toast being 'Success to the Trig'.

William Turner's dramatic engraving of Loch Coruisk, set deep in the Cuillin.

The 'perpendicular ridges' on Skye. This an early 20th century photograph of the West Ridge on Sgurr Nan Gillean.

Osgood Mackenzie's uncle remembered Colby's visit – the family had offered him hospitality, Highland custom requiring that so distinguished a visitor should not have to stay at an inn. He speaks with great affection of Colby, stating that 'he retired afterwards to Kerrysdale, and seemed to be more peaceful and happy than anyone I ever knew' (Mackenzie, 1949).

CHAPTER THREE

THE EARLY TRAVELLERS

I now have the pleasure of going where nobody
goes, and seeing what nobody sees.

(Dr Johnson)

To the average European, a trip to the Highlands would not have been an obvious destination in the years following Culloden. When Boswell mentioned the possibility of a tour to the north of Scotland to his friend the French philosopher Voltaire, 'he looked at me, as if I had talked of going to the North Pole, and said "You do not insist on my accompanying you?" "No, Sir." "Then I am very willing you should go"' (Boswell, 1984).

To many, it was frightening: not just the terrain, the high mountains, the lack of roads and maps, and the reputed weather, but also the people, with their distinctive garb, their foreign language, and alien customs. In spite of the defeat by the Duke of Cumberland, their reputation for wild aggression must have lingered into the 1750s. The government admitted to a 'total neglect of the affairs of that country' (*The London Magazine*, February 1747), while the ordinary citizens, whether in Edinburgh or London, gave little thought to their fellow countrymen in the far north.

Gradually, though, the events of 1745–1746 began to change attitudes. Those to the south had been shocked into acknowledging the existence of these northern regions, and had taken means to compel them, however unwillingly, to submit to British governance. With this acknowledgement came a new interest in the area. What really lay beyond those mountain ranges? The scientists – geologists, botanists – were naturally attracted, sensing a virgin territory, but there were others of intelligent and enquiring minds who were bold enough to consider the difficult journey into that unexplored region. 'Scotland is full of curious things', wrote the Reverend Richard Pococke to the Foreign Secretary of the Royal Society, who was planning a trip to the Western Isles, 'and as they have not been much searched into … you will … be a Columbus in New discoveries'. Pococke himself enthusiastically donned the mantle of Columbus, making three trips to Scotland – the longest, his last, in 1760. He

"The Right Rev. Richard Pococke D.D. Bishop of Meath." A portrait taken from the account of his tours published in 1887.

Thomas Pennant, a portrait after Gainsborough, published in The European Magazine.

hoped, in the far north 'to meet with something curious, at least in relation to … customs and manners' (Pococke, 1887).

So it was too, with Johnson and Boswell: 'Our business was with life and manners' (Johnson, 1984). Their accounts of their journey to the Western Isles in 1773 form some of the most famous travel writing of all time, and they showed courage in the face of many who doubted the wisdom of such a venture. Clearly Boswell's wife had her reservations, for she was informed that 'We do not go there as to a paradise. We go to see something different from what we are accustomed to see' (Quoted in introduction, Johnson, 1984). Johnson expected it to be a dangerous expedition, but Boswell was confident of their safety, and persuaded the Doctor to leave his pistols behind in Edinburgh.

The two men were encouraged by the account of another visitor to Scotland, that of Thomas Pennant in 1769. Johnson was impressed. 'The best traveller I ever read: he observes more things than anyone else does' (Boswell, 1907). Pennant undertook another journey in 1772, the year before that of Johnson and Boswell, but his account was not published until after they had returned. His first journey had not taken him as far north as he would have liked – he felt there was more to be seen. 'This journey was undertaken in the summer of

1772, in order to render more complete, my preceding tour; and to allay that species of restlessness that infects many minds, on leaving any attempt unfinished' (Advertisement to *The Tour…1772*, Pennant, 1998). He added in the *Scots Magazine* just before leaving that 'my sole objects are, my own improvement, and the true knowledge of your country, hitherto misrepresented'. As well as being a traveller and an antiquarian, Pennant was also a keen naturalist. On his second journey he was accompanied by the botanist the Reverend John Lightfoot, as well as the Reverend John Stuart of Killin who knew something of Scottish ways and manners, and the artist Moses Griffiths, whose wonderful engravings enhance both volumes of the Tours.

Another visitor who dared to venture to the far north, the Reverend Charles Cordiner, wrote his account in 1780 in the form of letters to Thomas Pennant. One site that attracted all these antiquarians to the Sutherland wastelands was Dun Dornadilla (Dun Dornaigil), the ruins of a broch that stands near Loch Hope. Pococke and his party trudged all the way up Strathmore to see it, while Cordiner gave a vivid account of his ascent of Ben Hope in order to get there. Pennant quotes an anonymous author writing in the *Edinburgh Magazine*, who called it 'the greatest piece of antiquity in this Island' (Pennant, 1998). Looking at it now,

Dun Dornadilla: "Inside View of a Round Castle." Pococke's sketch taken from his account published in 1887.

this might seem quite surprising. It is only 20 feet high, and although it has a fine entrance which is crowned with an enormous triangular lintel, the inside is inaccessible, since it is full of stones that have fallen from the structure into the centre over the years. Clearly, it was a more imposing edifice in the eighteenth century. It stood at least 30 feet high, and it was possible to enter, as Pococke's sketches show. Alexander Bryce had observed its double walls and a spiral staircase when researching his map of 1744. No one knew what it was. Some said a hunting lodge or a fortress, others a temple, or even a tomb. The unknown gave it added appeal. Even today, its purpose is not entirely understood; it was probably some sort of defensive shelter for the local farmers, perhaps more a status symbol than a fortress. It is thought to be over 2,000 years old.

Johnson and Boswell did not venture as far north as these other visitors. They turned to the west at Inverness and made their way to Bernera, thence to the Western Isles. In some ways they were a little disappointed. 'We came thither too late to see what we expected, a people of peculiar appearance, and a system of antiquated life… Of what they had before the late conquest of their country, there remains only their language and their poverty… Such is the effect of the late regulations, that a longer journey than to the Highlands must be taken

"Dr Johnson in the Highlands". A painting by the Marquess of Lorne, published in The Graphic, *1893.*

by him whose curiosity pants for savage virtues and barbarous grandeur' (Johnson, 1984). Their visit, nevertheless, made a deep impression on them. At the end of the tour, Johnson told Boswell that it had been 'the pleasantest part of his life' (Boswell, 1984). A remarkable comment given that we learn that they faced some dreadful inns, variable food, poor or no roads, and travel discomforts that must have been acute to a man of Johnson's size. They seemed to take little consolation in the scenery that now takes our breath away – for them it was 'this wide extent of hopeless sterility' (Johnson, 1984). On one particularly wet day, Johnson asked, 'Who *can* like the Highlands?' But he did add, 'I like the inhabitants very well' (Boswell, 1984).

It would be fair to say that most travellers to these regions, including those of the nineteenth century, liked the inhabitants very well. There are two adjectives that recur in descriptions of the Highland population; 'hospitable' and 'indolent'. Of the latter, more later: of the former, examples abound. 'We scarce passed a farm', writes Pennant, 'but the good woman, long before our approach, sallied out and stood on the road side, holding out to us a bowl of milk or whey' (1998). John Ramsay of Ochtertyre records meeting a farmer who told him 'that he allowed nobody to pass his house without bread and cheese and a drink of milk or whey; but if a gentleman or minister came into the glen, he killed a lamb or a kid' (Fyfe, 1942). Whilst traversing the very north of the country, Cordiner thought that 'the hospitality of these coasts appears no less romantic than the scenery. The being a stranger seems to be a title to every office of friendship, and to the most distinguished marks of attention and civility' (1780). Some observers, a little cynically, found reasons for this generosity. The Reverend Thomas Munro in the 1845 *Statistical Account* thought it stemmed from 'a love of ostentation, and a

"The Highland Breakfast".
From a painting by John Phillip.

spirit of independence which has sometimes exercised the wit of their more refined neigh-bours in the South, under the name "Highland Pride". Sir John Dalrymple, on the other hand, suggested fear was behind all Highland politeness. 'Every man wore a dirk…and no man dared to be rude to his neighbour, lest he should receive the skene dhu in his wame [stomach]' (Quoted in MacCulloch, III, 1824). But these remote, isolated communities could not hope to survive without a spirit of generosity to one another, and it is significant that in the more populated town of Dingwall, it was noted that 'the lower order of people is not remarkable for any extraordinary degree of hospitality' (Sinclair, 1799).

Johnson was hoping for 'people of peculiar appearance' (Johnson, 1984). Perhaps he did not have in mind the lad with *erectis auribus* (protruding ears), of whom Pennant remarked, 'His ears had never been swaddled down, and they stood out as nature ordained; and I dare say his sense of hearing was more acute by this liberty' (Pennant, 1998). Johnson was thinking more of the Highland garb, which had been forbidden since 1746. Had he travelled further north, he might have seen more of it. He had already noticed on arrival in Inverness that 'the appearance of life began to alter. I had seen a few women with plaids at Aberdeen; but at Inverness the Highland manners are common' (Johnson, 1984). According to Colonel David Stewart, the plaid was a piece of tartan which measured '2 yards in breadth and 4 yards in length' (see colour section, page ii). There was sufficient material to be 'confined by a belt, buckled tight round the body, [and] while the lower part came down to the knees, the other was drawn up and adjusted to the left shoulder, leaving the right arm uncovered, and at full liberty' (Stewart, 1822). It was nothing if not an adaptable piece of clothing: in wet weather it could be draped over both shoulders. Astonishingly, Stewart continues that sometimes at night, the Highlander 'would dip the plaid in water, and, wrapping himself up in it when moistened, lie down on the heath. The plaid thus swelled with moisture was supposed to resist the wind, so that the exhalation from the body during sleep might surround the wearer with an atmosphere of warm vapour' (ibid.). It was a garb that these northern dwellers found ideal, but Alexandre de la Rochefoucauld was offended by it. The sight of bare flesh visible below the kilt 'seems very indecent to us'.

At this distance in time, it is hard to grasp exactly what tartan represented as a concept, for the distinctive patterns associated with each clan, that we know so well today, are a nineteenth-century invention. It seems as though the cloth, that was woven at home, was coloured using whatever dyes were obtainable in the vicinity. *Certaine Matters concerning Scotland* which was published in 1603 says that Highlanders 'delight in marbled cloths, especially that have long stripes of sundrie colours; they love chiefly purple and blue … but for the most part now, they are brown, most near to the colour of hadder [heather], to the effect when they lye among the hadders, the bright colour of their plaids shall not bewray [*sic*] them' (Stewart, 1822). The author adds that 'they suffer the most cruel tempests that blow in the open fields, in such sort, that in a night of snow they sleep sound'.

What surprised many visitors to Scotland was the number of people who wore no shoes. At Inverness, Johnson observed 'Tall boys…run without them in the streets and in the islands' (1775). By 1820, the geologist John MacCulloch noted that 'it is extremely rare

Two highland lasses (possibly mother and daughter) in an original painting by the English artist Paul Falconer Poole.

Bergère Écossaise. A somewhat idealised French image of a Scottish Shepherdess.

to see a man barefooted…but it is equally rare to see a woman with shoes, except when in full dress, on Sundays'. He adds that 'the children of both sexes are bare-legged even to an advanced age, not only among the poorer classes, but also in families of condition' (MacCulloch, I, 1824).

Pococke was impressed by the 'well-bodied men of great activity, … [who] go the Highland trot with wonderful expedition' (1887). He found that the post-runners thought nothing of covering 50 English miles in one day, and even the ten-year-old postboy had done the journey from Ratter (Rattar) to Thurso and back by 11 in the morning – said by Pococke to be a journey of 16 miles in all, but now, in English miles, measured at 24! Alexandre de la Rochefoucauld thought the males generally strong and handsome, ideally suited to a career in the army. The women, therefore, were left to do much of the work. 'The women are strongly built and work well, usually in rather dirty work, and they are badly dressed'. He saw them loading manure into carts, using their bare hands. However, north of Aberdeen he found the women 'extremely pretty, and much more "my type". They're not like the dirty and ugly creatures we found in the Edinburgh neighbourhood'. Even Johnson at one point appeared to succumb to female charms, stating 'the girls of the Highlands are

"Carrying Ferns" (top left), and "Carrying Home Peat" (top right), from the 1900 edition of McIan's Highlanders at Home.

Left: Girl with a creel on Skye. A photograph (c.1880) by George Washington Wilson. She is collecting seaweed to use as fertiliser.

all gentlewomen'. He took a shine to his host's daughter at Anoch. 'I presented her with a book, which I happened to have about me, and should not be pleased to think that she forgets me' (Johnson, 1984). Everyone was keen to learn what book he had selected as being suitable for a young lady on an occasion like this, and was more than a little disappointed to discover it was Cocker's *Arithmetick*! Poor Doctor Johnson! According to Boswell, the incident provided later much mirth amongst 'the ladies', at which Johnson would become 'a little angry' (Boswell, 1984). He would assure them that it was the *only* book which he happened to have about his person.

On the mainland, Johnson and Boswell never strayed far from the Scottish road system, such as it was then. The most difficult stretch of their journey was from Inverness to Bernera, on one of General Wade's famed military roads. Wade had been part of the force that had quelled the first Jacobite rebellion in 1715. The British Government, aware for the first time that its hold on the Highlands was minimal, had sent the General back to Scotland in 1724 to suggest improvements that would contain a similar uprising. He recommended improving old, and building new, barracks at key strategic points, linking them with a series of roads to enable fast troop deployment. The barracks were located at Bernera and Fort William on the west coast, Ruthven, Fort Augustus, and Inverness (Fort George). Over the next 12 years, using parties of up to 300 soldiers, Wade supervised the construction of 240 miles of road. The only major engineering construction on the project was the handsome bridge that still spans the Tay at Aberfeldy, but the roads themselves are also a fine achievement, often crossing the most rugged and difficult terrain. The road along Loch Ness, which had involved blasting out a lot of rock, was particularly admired. As Johnson observed, it was 'levelled with great labour and exactness' (Johnson, 1984). By the end of the project, which was completed by Major William Caulfield after Wade had moved on, over 800 miles of

"Ruthven Castle", an illustration taken from Charles Cordiner's Remarkable Ruins and Romantic Prospects of North Britain…., *the engraving by Peter Mazell. The Castle was reduced to a ruin in 1746 by the Highland Army.*

"New Fort George". An engraving by Charles Cordiner.

Road-builders shifting boulders. A pen and watercolour sketch by
Paul Sandby. Image courtesy of the National Galleries of Scotland.

road had been constructed. Perhaps Caulfield's greatest legacy, though, was the verse he composed for which (as Joseph Mitchell noted in 1883) 'few poets for a single couplet have obtained such immortality' (Mitchell, 1971):

> Had you seen these roads before they were made
> You would lift up your hands and bless General Wade.

Not everyone blessed the General. To begin with, the locals did not want the new roads. Peter Graham of Rudivous 'saw no use of them but to let burghers and red-coats into the Highlands' (Ramsay Notebooks, in Fyfe, 1942). Clan chiefs feared that it left the country more open to invasion, and even dismissed bridges, which 'rendered their people effeminate and less fit to undertake the crossing of rivers' (Mitchell, 1971), regardless of the fact that every year a number of people drowned when trying to ford these rivers in spate. At Glencoe, Johnson (1984) noticed that the milestones had been removed, 'resolved, they said, "to have no *new* miles"' (English miles, at 1,760 yards, which were shorter than the Scottish mile of 2,400 yards). The 1799 *Statistical Account* summed up the Highland attitude to good communications with the maxim, 'The more inaccessible, the more secure'.

However, the roads had not been constructed for the convenience of the locals, but rather, for troop movements. As a result, they could be quite narrow, and were certainly not designed for wheeled carriages. The surveying was minimal, with roads following as straight a line as possible, no matter what obstacle lay ahead. The Honourable Mrs Sarah Murray, whose *Useful Guide to the Beauties of Scotland* was published in 1799, wrote of Wade that 'his military roads generally go up and down mountains, [he] never dreaming that he could wind round the bases of them.' She was alarmed at Loch Ness where the road had 'a precipice of perhaps eighty or a hundred feet perpendicular; [which is] no security whatever … should the horses there take fright.' Johnson found the road over Ratiken (Ratagan) so steep that his horse 'staggered a little', and confesses it was 'the only moment of my journey, in which I thought myself endangered' (1984). The roads at times were just too steep and too narrow; the locals preferred the old tracks, and by 1790 Wade's highways were in a poor state of repair.

But at least there were roads: north of Inverness, there were none. In his report on the Annexed Estates, Captain John Forbes struggled when surveying those of Lovat and Cromarty in 1755: 'The roads to and through this country may be reckoned amongst the worst in the Highlands of Scotland, being mountainous, rocky, and full of stones, and no bridges upon the rivers, so that nothing but necessity makes strangers resort here and for a great part of the year it is almost inaccessible.' People spoke of 'drover's roads', but in the far north there was little restriction of movement and passage. A 'top man' went ahead to choose the route depending on the prevailing conditions. The result was that there were fewer well-defined tracks than might be expected. Most of the cattle in Sutherland were taken east to Helmsdale Bay, thence south, crossing the Dornoch Firth at Meikle. The animals were

"Loch Leven at Ballachulish Ferry". Drovers arriving at Glencoe
on their way south. An engraving after Joseph Adam.

"The Height of Ambition". An engraving after Jacob Thompson,
showing transport using a sledge, rather than a wheeled vehicle.

Highland Ponies. After a painting by Rosa Bonheur, 1867.

required to swim over these stretches of water. If they did so willingly, it was seen as a good omen for the sale. There was also a route down the west coast, crossing to the Muir of Ord in the east by Glen Cannich, Glen Affric, or Strath Garve. Wade did follow some of the old drover tracks when making his roads further south.

With no maps available – simply undefined tracks and paths – a guide was essential for visitors. Cordiner describes how at Armisdale (Armadale), travel became 'extremely difficult. For several miles the best road which the guides could take was in the channel of a rivulet, and its bed was far from being smooth' (1780). There was no way that an outsider could have known that this channel was the road.

Any sort of wheeled vehicle was out of the question – there were none to be found, anyway. Even the type of horse was limited. As Cordiner's party entered Sutherland from Caithness, they had to exchange their large horses for the smaller local breed:

> Large horses cannot take that route [to Ross-shire via Sutherland], not only on account of the exceeding roughness of the rocky heaths; the difficulty of the paths among the hills, where climbing is often necessary, and the dangerous nature of the swamps and morassy grounds: but, as it is not practical to carry corn and hay into the wilds; the finding of good grass being extremely precarious, and provender of any kind very difficult to be obtained; all they who wish to penetrate into the more remote and desert districts of Strathnaver, must be

furnished with the hardy ponies of the country; a breed I believe originally from the Orcades [Orkney]. These being accustomed to climb among rocks; to jump between hillocks among bogs; to feed on birch-leaves, or any green stuff that grows among the hills, are the only proper horses for the journey. I was favoured with a couple of these, and guides who could speak both the Erse and English languages (Cordiner, 1780).

Even Dr Johnson had to mount such beasts. 'As Dr Johnson was a great weight', notes Boswell, 'the two guides agreed that he should ride the horses alternately' (Boswell, 1984). The Doctor himself, seated upon a small pony, admitted that 'had there been many spectators, [I] should have been somewhat ashamed of my figure in the march... A bulky man upon one of their backs makes a very disproportionate appearance' (Johnson, 1984).

Riding through these northern parts, no one was more aware of the changes that were taking place than Johnson and Boswell. Those changes were being driven from the top. 'The chiefs, divested of their prerogatives, necessarily turned their thoughts to the improvement of their revenues ... as they have less homage' (Johnson, 1984) was how Johnson succinctly put it. These leaders, who for centuries had held a form of feudal control over their subjects, found themselves emasculated after Culloden. Their tartan finery and weapons were taken from them, and their powers of jurisdiction removed. With this sudden and immediate loss of privilege and standing, it was inevitable that they would look elsewhere for compensation. They found it in the pursuit of wealth, and sheep.

Unfortunately, this was not a society that would respond well to change. It had always been poor, materially speaking, but it had functioned. The tenants respected the Lairds, put in the hours of labour as required, and went to war if necessary. In return, their chiefs looked after them, finding extra meal for them in times of famine, and looking the other way when they got hopelessly behind with their rent. It was only when their status as chiefs was taken away that the system began to seem irksome to them.

This was the situation in the Highlands that these travellers found as they made their way slowly further and further north. MacCulloch called the inhabitants 'a docile people' (III,1824), who were generous of spirit, contented with so little, but resistant to change, and confused by what was going on. The indolence, of which they are so often accused, had been characteristic of a system that had worked for centuries. Now, to these outsiders, it seemed almost wilful, and distressing. This is what Pennant found in Assynt:

This tract seems the residence of sloth; the people almost torpid with idleness, and most wretched: their hovels most miserable, made of poles wattled and covered with thin sods. There is not corn raised sufficient to supply half the wants of the inhabitants: climate conspires with indolence to make matters worse; yet there is much improvable land here in a state of nature: but till famine pinches they will not bestir themselves: they are content with little

Three generations of crofters. An anonymous 19th century photograph.

at present, and are thoughtless of futurity… Dispirited and driven to despair by bad management, crowds were now passing, emaciated with hunger, to the Eastern coast, on the report of a ship being there loaden with meal. Numbers of the miserables of this country

"The Prophecy of Famine". The frontispiece to a Scots Pastoral Poem by C. Churchill, published in 1763.

were now migrating: they wandered in a state of desperation; too poor to pay, they madly sell themselves, for their passage, preferring a temporary bondage in a strange land, to starving for life in their native soil (Pennant, 1998).

It would be a mistake to think that scenes like this were widespread throughout the Highlands at this point. The mass expulsions of the population from the inland glens that were a feature of the early 1800s had not yet begun. All the visitors comment on the poverty of the people, but the inhabitants had lived like that for years, and seemed almost content with it. The most striking evidence of this poverty was their housing. Near Inverness, Alexandre de la Rochefoucauld was alarmed by 'the dwellings…made of earth, with clods of earth for roofing; the bleakest picture of poverty, these hovels are falling into ruin, and so low that one can scarcely scramble in through the entrance' (Scarfe, 2001). The owners of such houses would think nothing of sharing them with their livestock, as Sarah Murray observed. The farmer slept separated from his animals only by 'a very partial partition; he delights to sleep

"A Cottage in Ilay". An engraving by Moses Griffiths, from Pennant's Tour, 1772.

Some houses remained unmodernised throughout the 19th century.
This an anonymous photograph titled "A Highland Cottage".

thus close to the byar [*sic*], that he may lie and see, and hear, his beasts eat. Another pretty fashion among them (it is universal), [is that] their dunghill is close to the door of the house, or hut: let the spot about it be ever so lovely, to them their sweet mixen [manure heap] is their choicest, their chief object… What a perverse inclination for nastiness!' (Murray, 1799). Later visitors found that the situation had not changed at all. On Islay, where he supped one evening, MacCulloch was troubled by long cow's tails whisking away behind him. 'The grog was overset, and the candle extinguished' (MacCulloch, III, 1824). At Loch Sloy in 1803, James Hogg found the situation positively dangerous: he was staying with ten others in a hut with no furniture. They all slept on 'the same floor with four or five cows, and as many dogs, the hens preferring the joists above us. During the night the cattle broke loose, if they were at all bound, and came snuffing and smelling about our couch, which terrified me exceedingly'. The dogs were encouraged to keep the cows at bay, which was just as well as 'they scarcely … missed trampling to death some of the children, who were lying scattered on the floor' (Hogg, 1888). It was not only in the houses of the poor that the accommodation left something to be desired. 'We were driven once', writes Dr Johnson, 'by missing a passage, to the hut of a gentleman, where, after a very liberal supper, when I was conducted to my chamber, I found an elegant bed of Indian cotton, spread with fine sheets. The accommodation was flattering; I undressed myself, and felt my feet in a mire. The bed stood upon the bare earth, which a long course of rain had softened to a puddle' (Johnson, 1984).

Charles Cordiner made his way intrepidly right up to the far north-west of Sutherland. His map told him he was in Dirry-more Forest, to the north of Loch Shin, but he was surprised to find 'scarce the vestige of a tree discoverable in it' (Cordiner, 1780). He was forced to seek shelter in a 'keeper's booth, the most wretched hovel imaginable; with difficulty to be crawled into, and one could not stand upright when within … Yet here was straw for a bed, a bottle of milk, and some pieces of bread. It is possible men may be contented with such spare accommodation, with such hard and scanty fare. I strove in that cot to find shelter from a heavy rain, but soon found such confinement seem[ed] worse than an open exposure to the severest weather' (ibid.).

However, there were cultivated oases upon which to stumble unexpectedly within these 'dismal wilds and mountain fastnesses' (Lawson, 1842). One such was at Strath Hallad-dale [Strath Halladale] where Cordiner came across 'a rich but narrow valley' producing 'fine natural grass'. There, he found a garden, the appearance of which 'was unexpectedly pleasing … the borders decked with [a] variety of the richest flowers, plenty of wall fruits; apples, pears, plums, cherries, which are often as early ripe as at Edinburgh; beds of melons and cucumbers; and whatever can give variety, or grace the entertainments of the table' (Cordiner, 1780). Another impressive garden was at Lord Reay's house in Tongue, from where the chief had watched the *Hazard* founder. Pococke visited it on his way along the north coast, heading east, when he noticed 'a handsome terrace and bowling green between the house and the bay, and a kitchen garden behind the house planted with all kinds of fruit except peaches, apricocks [*sic*], and plumbs [*sic*]. Cherries and apples are planted against the walls; and in the middle of the kitchen garden is a pillar entirely covered with dials [a sundial that still stands in the grounds] … There are large plantations of wichelm, ash, sycamore, and some quicken or mountain ash' – these, no doubt, were a welcome sight in a landscape so denuded of trees (Pococke, 1887).

Cordiner does not hide the fact that the travelling was arduous and difficult – in fact he revels in it. He enjoyed the scenery at Tongue, and 'had one only to admire this so various and noble scenery, and not travel through the horrid paths … it would be extremely delightful.' Alas, the guides were warning of a difficult route they would be following on the morrow, 'but the hope of seeing the celebrated Dun of Dornadilla, banishes every other care, and animates the thoughts of the journey.' The next day, the guides confirmed that because of the 'numberless pools, and mossy ground softened into bog by perpetual streams from the hills', it would be necessary to go over, rather than round, Ben Hope, and for this another guide would be needed. It seems unlikely that they crossed the actual summit – there was certainly no need to – but a fair bit of climbing would have been unavoidable. Cordiner's pony 'wandered through with much seeming unconcern', unlike the rider, who feared that 'the danger of being thrown, on such road, was considerable'. He thought about dismounting, but was advised to stay on for as long as possible. Eventually he had to walk, leaving the horses free to follow as they would, which they did 'with great dexterity'. He indulges in descriptions of 'vast chasms' and places 'where one might plunge to an immeasurable depth' and at length they found themselves in cloud. With the temperature falling, they 'quickened [their] pace, to get beyond the highest part of [their] route, and soon found the difficulty of surmounting Ben Hope was over'. There below them lay Dun Dornadilla, and with it 'rising corn…[which] mingled a rural softness with the vast wildness of the rest of the prospect'. As for the broch, it was in 'a situation distinguishedly romantic, magnificently wild' (Cordiner, 1780).

"Dun Dornadilla", with Ben Hope in the background.
By Charles Cordiner, from his 1780 account.

"Loch Naver", after a painting by John Fleming, from Swan's Lakes of Scotland
published in 1834. Ben Klibreck is in the distance. Views of mountains like
Ben Klibreck and Ben Loyal are very scarce before the 20th century.

Pococke approached the same spot from the south, and makes less of the trials of the journey. At Lairg, he made the acquaintance of the Minister, who accompanied him to Ben Klibreck – already a considerable distance. They enquired about accommodation but 'as it did not please us, we went on'. One wonders what was offered to them: possibly shielings – huts the farmers used in the summer months, of the type that so repulsed Cordiner in Dirry-More. The weather was poor. 'Here, it was like the month of November' (it was in fact 22 June), but they sat down and took their refreshment, sharing it with some boys who were in the vicinity looking after cattle. As they were leaving, the boys' mother appeared some way off. 'She carried a piggin of cream, which was warm whey. She drank to us, and we took it round and tasted of the whey' (Pococke, 1887). Such were the simple delights of summer travel in the Highlands.

Pococke must have been excellent company: 'a man of mild manners and primitive simplicity' (Pococke, 1887). He clearly enjoyed his food, even in these northern regions, whether it consisted of the ears of a calf toasted on bread, or a pot-baked loaf, both of which he sampled when staying with Mr Murray of Pennyland, near Thurso. 'Stop, my Lord!' joked his host, 'Else your Lordship will raise a Famine in ye Country.' Known as 'Pococke the traveller' (ibid.), he had spent five years in the Middle East between 1737 and 1742, and this

A shepherd at his shieling in the hills. A photograph by Charles Reid (c.1890). Reid was born in Turriff, Aberdeenshire in 1838, and specialised in photographing livestock and wildlife.

A New & Accurate Map of that part of Great Britain called Scotland *by Thomas Bowen, c.1770. Bryce in his map of 1744 had drawn attention to 'The Moan (Moine), a great Morass.' Mapmakers like Bowen would have learnt of the difficulties such terrain presented from accounts like that of Pococke and Knox (see chapter 4) and made sure it was marked graphically, as here. However, Bowen also suggests a road along the north coast to Durness, but that did not exist until 1830.*

was his third trip to Scotland. From Dun Dornadilla, he continued all the way up to Cape Wrath, where, always keen to remind people of his Middle Eastern adventures, he compared the granite to that of which the statues of Memnon were made. The weather had maybe improved as he was able to see 'a great part of the Isle of Lewis, and the Isle of Ronon [North Rona]'. He saw boys fishing for Cudines (coalfish), and a pair of eagles which had nested on two high rocks of granite. Later they found the fawn of a red deer which had been killed by eagles: 'the hinde on this occasion runs about, and stamps with her foot and cries terribly. But the eagles will, they say, kill a hart by seizing …[her] about the neck and fluttering their wings in …[her] eyes'. He was told that there were a large number of adders in the vicinity, which led him to muse on whether it was true that goats eat snakes. 'It was confirmed to me here in such a manner, that I could not withhold my belief of it, and, 'tis added, that they make a great noise when they kill them'. In Daniel Kemp's *Biographical Sketch* of Pococke (1887), he mentions various Gaelic sources that refer to goats eating snakes – it was clearly widely believed in the Highlands.

He came across a herdsman who was living at Kerwick Bay (Kearvaig), possibly the great Sutherland bard Rob Donn who was, at one time, Lord Reay's shepherd there. Pococke certainly met the minister of Durness, for the Reverend Macdonald mentions it in his diary. 'Most of the week taken up with a conspicuous stranger, Dr Pococke … He seems to be curious, ingenious and judicious, and I hope our country may not be the worse of his visit, which has probably rubbed off prejudices *hinc inde* [as a result]' (Grimble, 1999). The minister's diary reveals particular details of life in this remote part of the world, not least the difficulties of travel, whether attempting to attend a funeral in Thurso or simply trying to reach Tongue across the Moine. The latter tract was a notoriously bad stretch of boggy ground that was only relieved by the making of a road in the 1830s. In June 1749, the Reverend records 'On the Monday I was obliged to pass over the Moine, a tedious, fatiguing moss of five miles on the way, in worse weather than I had in December.' Then, in 1755, 'in the form of a boisterous wind I came over the Moine and preached this day at the Meeting House here in much weakness of bodye [and] mind, occasioned partly by the toil and weariness of the journey.' Pococke was escorted over the Moine by a Mr Forbes, noting that it is 'a morassy country, impassable, except to their little bog horses.' This is not the last we will hear of the Moine.

Pococke continued along the north coast, enjoying a fine view of Ben Hope, though he thought it 'did not appear so beautiful with its pointed top as when it was covered with cloud.' Near Tongue, he remarked on the ruins of Castle Varrich, perched on a hill overlooking the town, and also the less obvious 'foundation of a round castle on an eminence, now entirely destroyed. To the south is a fine craggy mountain called Ben Loyal.' The so-called round castle was probably the broch, Dun Mhaigh, which is located at the end of the Kyle, and does indeed have fine views of the 'Queen of Scottish Mountains'. At Farr, he was entertained with cake and a glass of malaga (a sweet, fortified wine) by Captain Mackay, 'a half-pay officer of Holland', and found himself a little alarmed as they rode along the wide bay which 'consists of soft sand…not without some apprehensions to a stranger, tho' all was safe'.

Thomas Pennant and his party were unable to reach the far north coast, having been informed 'that the way was impassable for horses … and that even an highland foot-messenger must avoid part of the hills by crossing an arm of the sea' (Pennant, 1998). Pennant was at his most ecstatic when staying at Dundonnell:

> A spot equalized by few in picturesque and magnificent scenery … To the west is a view where the aweful, or rather the horrible predominates. A chain of rocky mountains, some conoid, but united by links of a height equal to most in *North Britain*, with sides dark, deep, and precipitous, with summits broken, sharp, serrated and spiring into all terrific forms; with snowy glaciers lodged in the deep shaded apertures. These crags are called *Sgur-fein*, or hills of wine; they rather merit the title of *Sgur-shain* or rocks of wind; for here *Aeolus* may be said to make his residence, and be ever employed in fabricating blasts, squalls and hurricanes, which he scatters with no sparing hand over the subjacent vales and lochs.

The mountain that had so caught Pennant's attention is now known as An Teallach. It boasts one of the finest ridges on the Scottish mainland.

"Dundonnell." After Moses Griffiths, published in Pennant's account, 1774. The distinctive ridge of An Teallach can be clearly seen.

Previously, Pennant, with his companions, had struggled on as far north as they could go, perched on 'shoeless little steeds', reaching Ledbeg in Assynt where they obtained 'quarters, and rough hospitality, from a gigantic and awful landlady'. Pennant would, therefore, have passed under Knockan Crag, a key site that will feature later in the geological controversy which brought this remote region to the attention of the world. It is somehow fitting that he should have passed that way: he who had described the area as a 'country that seemed to have been so torn and convulsed: the shock, whenever it happened, shook off all that vegetates' (Pennant, 1998). Here, indeed, he was in the shock zone.

And what of Mrs Sarah Murray? She was last seen far to the south, and she never left the comfort of the roads. After reaching Inverness, she headed back south by Loch Ness. At Aviemore the 'inn was within sight when I came down to the side of the Spey; and my heart jumped at the idea of passing the night in a spot so grateful to my sensations, because nature there shines in its natural garb, and in high beauty: but no sooner had I put my foot within the walls of that horrible house, than my heart sunk [*sic*]; and I was glad to escape from its filth and smoke very early the next morning' (Murray, 1799). Having declined breakfast at this establishment, she stopped at 'a small house, eight miles further on the new road to Dulsie Bridge … I got a comfortable meal in the chaise, having provided tea, sugar, bread and butter, tea-pot, &c. so that I wanted only boiling water and milk, which I got, extremely good, from the cottage'. These packed meals were to be, perhaps quite sensibly, her solution to the rather variable highland cuisine on offer, but without wishing to belittle the achievements of an intrepid female travelling on her own in the Highlands, she comes across as a slightly aloof figure, observing all from the safety of her chaise. However, she does leave us with one particularly moving vignette of a country on the verge of change and transformation, with which to close both this chapter, and this century:

I had observed no beggars in the Highlands, till I came upon the Moor between High Bridge and Fort William; but there, at the sound of the carriage, came bounding like fauns, through the dub and the lare (mire and bog), swarms of half-naked boys and girls, muttering Galic [*sic*]. Having no half-pence, I shook my head, and made every sign I could think of to make them understand I had nothing for them; but notwithstanding, one fly of a girl kept skimming over everything in her way, by the side of the carriage, for at least two miles; I screaming, 'tomorrow I will give you something'. Whether she became weary, or conceived what I meant, I cannot say; but at length she took a different direction and bounded away through bog and heath, to a hut on a dismal looking swamp,

"Near Tarbert, Loch Fyne". A sketch from The Graphic, 1875, with children gambolling enthusiastically beside a coach.

at some distance. On the morrow, the rattle of the wheels again brought forth a swarm, and my skipping lass amongst them; I had not forgotten her; but all Maryburgh could not furnish me with six-penny worth of half-pence. The girl bounded before me smiling; and seemed to express by her countenance, that tomorrow was come, and that she claimed my promise. On a steep rise she came close to the window of the chaise; she did not speak, but she looked in my face so expressively, that out came a silver six-pence from my purse, and I threw it before her. She stooped to pick it up, expecting, I suppose, a half-penny; but no sooner did her eye catch the white metal but she jumped a full yard from the ground, uttering such a scream of joy and surprize as startled me, and might have been heard at a great distance. She then quickly turned to her companion beggars, shewed the sixpence to them, and, with a smile of delight, bounced away towards the huts with an incredible swiftness. I never gave a six-pence with so much pleasure in my life; nor do I suppose one ever was received with more ecstasy (Murray, 1799).

CHAPTER FOUR

THE REFORMERS

It is therefore expected by the public, that every proprietor who is capable of looking forward, will take his stand in this great work of national improvement … the names of such proprietors who shall, with a liberal hand, come forward … will be engraved upon every Highland rock, and be recorded with applauses to the end of time.

(John Knox)

Following Culloden, the Act of Proscription stayed in force for nearly 40 years. It is not clear when, exactly, southern attitudes to the Highlands softened, and its reputation as an area rife with hostility and rebellion was transformed to one deserving of aid and pity, but on 28 May 1778, at the Spring Garden Coffee House, the Highland Society of London was formed by 'twenty-five Gentlemen, Natives of the Highlands of Scotland' with the intention that it 'might prove beneficial to that part of the kingdom' (Sinclair, 1813). Six years later, at a similar venue in the Scottish capital – Fortune Toutine's Tavern – the Highland Society of Edinburgh was founded on 9 February 1784. This latter had three aims in 1785: to enquire into the present state of the Highlands; to investigate how to improve conditions in the Highlands and ensure that the Government was prepared to see these 'beneficial purposes' through, and thirdly, to 'preserve the language, poetry, and music of the Highlands' (ibid.) – the very things that had been suppressed since Culloden. In 1787 the Edinburgh Society received a Royal Charter: clearly, attitudes were starting to change.

The first president of the Highland Society of London was Lieutenant-General Simon Fraser of Lovat. He was the son of the notorious Simon Fraser, 11th Lord Lovat, 'the Fox' also known as 'the most devious man in Scotland'. In the rebellions of 1715 and 1745, he had tried to back both sides. The Government eventually saw through this ploy, and he was forced to flee for his life after Culloden. In due course, he was traced to Morar where he was found hiding in a tree, taken back to London, and executed on 9 April 1747. He has the distinction of being the last man to be beheaded in this country. Perhaps he took, at the very end, some wry satisfaction in the fact that a stand erected so that people could witness the spectacle, collapsed, killing 20.

The Fox's son, meanwhile, found himself in Glasgow under house arrest, but he received a full pardon in 1750. He then distinguished himself leading a force of Fraser Highlanders in North America under General Wolfe, and returned to Britain an establishment figure. He made sure that one of the aims of his Highland Society was 'keeping up the Martial spirit; and rewarding the gallant atchievements [sic] of Highland Corps' (Sinclair, 1813).

The Highland contribution to the British Army had been a little complicated for some time. Some Highland soldiers were actually based in the Netherlands, where since 1586 an Anglo-Dutch brigade, consisting of three English and three Scottish regiments had been formed. Technically this was a British unit on loan to the Dutch. They maintained their separate identity, wearing the red coats of the British Army, and marching to the 'Scots March'. The British Crown had the power to appoint the officers. We have already met one, Captain Mackay, offering cake and Malaga to Richard Pococke at Farr. Another such was Hugh Mackay, third son of Hugh Mackay of Scourie, who married a rich merchant's daughter in Amsterdam in 1673 and transferred to the Scots Brigade. He brought over 1,100 men from the Scottish regiments in Holland to fight in the Jacobite skirmishes of the 1690s, and was eventually killed fighting for the Dutch against the French at the Battle of Steinkirk in 1692.

The prowess of the Scottish soldier was legendary, not least his strength and fitness. In 1720, Defoe wrote, under the pseudonym Andrew Newport, that the cavalry of King Gustavus Adolphus 'always had some foot soldiers with them; and yet if the horse galloped, or pushed on ever so forward, the foot were as forward as they, and was an extraordinary advantage … These were those they called Highlanders; they would run on foot with their arms and all their accoutrements, keep very good order too, and keep pace with the horse, let them go at what rate they would'. Admittedly, Defoe's 'Memoirs of a Cavalier' turned out to be a work of fiction, but John Knox, too, had heard that 'a Highland regiment will outdo the Cavalry', though he added that 'in hot climates they frequently fall down and expire' (Knox, 1787). In the Netherlands, the soldiers that made up the Scots Brigade, as it was known, were considered the 'best in the Dutch State Army, the most courageous and trustworthy' (Migglebrink, 2003).

Gradually, it was realised in London that it would be to the country's advantage if these fine soldiers were brought fully into the British Army structure. A letter in the *London Chronicle* in December 1773 discussed the need, with the unsettled situation in North America, to 'avail ourselves of the assistance of that corps of British troops so being useless to its country, the Scotch Brigade, in the Dutch service; a corps long distinguished for its bravery and good behaviour, and which having been for some years past ill requited, and even very unjustly and harshly treated, by the Dutch, gives room to suppose, that a call to serve its native country will not at all be disagreeable to it'.

Matters came to a head in 1782 with the Netherlands siding with the French in America; the Dutch demanded the Highlanders swear an oath of loyalty, but most of them refused and

Above: "Bergschotten". A German engraving of a Scottish soldier, with his wife and hound (c.1835).

Left: "Chasseur Écossais". An original sketch of a Highlander, looking as much a soldier as a hunter. By the French artist Auguste Raffet.

consequently returned to Scotland. From this time onwards, the Highland regiments played an increasingly prominent role in the British Army, with the campaigns against Napoleon giving them every opportunity in which to shine.

Serious recruiting began to take place in the north, using fair means or foul. Donald Sage reported in his *Memorabilia Domestica* that in 1800, when trying to enlist men into the Sutherland Highlanders, General Wemyss procured a large amount of 'tobacco-twist and strong, black rapee snuff' with which to ingratiate himself. Alas, he had miscalculated: no one smoked, and they preferred the light-coloured snuff! None of it was taken – 'what became of the General's supply I know not'. With or without bribes, Highlanders, natural soldiers that they were, welcomed this employment opportunity and joined up willingly. The population figures for Kiltearn fell between 1791 and 1831: the *New Statistical Account* suggests that this was caused by 'the enlistment of numbers of the young men into the 42nd Regiment'. William Pitt the Elder, as Prime Minister at the time, was not slow to take the credit, claiming of Scottish recruitment 'I sought for merit wherever it was to be found. It was my boast that I was the first minister who looked for and found it in the mountains of the North' (Mitchell, 1970). It had been a while since such circles had associated merit with the Highlands.

Simon Lovat died in February 1782, thereby just missing the first major success of the Highland Society of London. On 1 July, the Government yielded to its request that the Act of Proscription be repealed. It seems that it had not been enforced for some time, but the Society was much encouraged by this success, and achieved a second, in 1784, when the forfeited estates in Scotland were restored. At last, there were signs of concern about the general level of poverty in the area, and the problems were beginning to be addressed. On 23 January 1784, James Anderson of Hermiston received a letter from the Secretary to the Honourable Board of Customs in Scotland informing him that a revenue-cutter had been placed at his disposal, so that he could carry out an investigation into the state of the Hebrides and western coasts of Scotland. Anderson was a prominent figure in the Scottish Enlightenment, an agriculturist who invented the two-horse Scotch Plough, and whose *Enquiry into the Nature of the Corn Laws* was said to have caught the eye of Karl Marx while he was forming his views on Capitalist Agriculture. He had been selected by the Government to carry out this survey because of his ability as an economic theorist. Unfortunately, as a result of what Anderson called 'a train of cross accidents' (1785), it was slow to get going, and he did not actually leave until 22 August, which was rather late in the season – not an ideal situation given the difficulties of travel he would face both on land and sea. He was not the only traveller who would regret finding himself traversing the country in a Scottish autumn, and he does admit that he was unable to cover as much ground as he had hoped. However, his report was published in 1785, and it is accompanied by a fine map. This shows a proposed road all the way up the west coast, and along the north coast, together with proposed canals at Crinan and along the Great Glen from Inverness to Fort William. Work on the Crinan Canal was underway within ten years, and on the Caledonian Canal some 30 years later, but it would be at least 60 years before all these roads were in place.

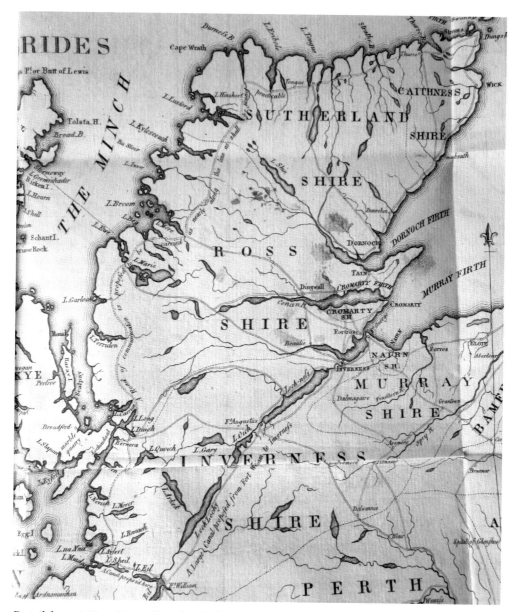

Detail from A New Map of Scotland, *the map that accompanied Anderson's Report in 1785. It shows the proposed road running all the way along the west and north coasts. It would be another 75 years before this route was finally constructed.*

The report is also noteworthy for a fierce criticism in the appendices of the charts of the Western Isles by Murdoch Mackenzie. Poor Mackenzie was a victim of his own success: his work in Orkney had been so complete and reliable that anything he printed was taken as accurate. In truth, his survey in the Western Isles was less thorough, and did contain errors. MacCulloch in the 1820s was still complaining about Mackenzie's 'blunder': he had omitted to point out a gap between Raasay and Flodda that is *dry* at low tide, nor had he mentioned 'a most dangerous sunk rock, lying exactly in the middle of this frequented passage [near Raasay], and not much more than a mile from the very house in which he resided [for] three weeks' (MacCulloch, III, 1824). The problem was that a mistake on a land map could be simply annoying; on a sea chart it could be life-threatening.

Meanwhile, at the same time as the Government was employing James Anderson, the Highland Society of London had instigated its own research. Its choice to lead this work was John Knox, a man of leisure who had made a comfortable fortune selling books in the Strand, London, and who had been making regular trips to Scotland since 1764. He described the Society as 'a convivial club, who met to enjoy themselves according to the customs of their country, to hear the bagpipe, [and] drink whisky out of the clam shell, &c.' (Knox, 1787), but they could turn their minds to serious matters as well. On 21 March 1786, at the Shakespeare Inn, Covent Garden, London, Knox was asked to draw up a paper on the subject of improvements in the Highlands. Within a week, Knox had composed his report, under the title *A discourse on the Expediency of Establishing Fishing Stations, or Small Towns in the Highlands of Scotland and the Hebride Isles*, and it was read

"A View of the British Fishery Off the S. Coast of Shetland". From the London Magazine, *1752*.

at a meeting in Richmond by the Earl of Breadalbane. His vision for Scotland included a more developed seafaring industry, with small, well-equipped towns all the way up the west, and along the north coast. The discourse was later presented to the Commons. To complicate matters, a new association was emerging at the same time, also based at the Shakespeare, called 'The British Society for Extending the Fisheries, and improving the Sea Coasts of the Kingdom.' It eventually became the British Fisheries Society, and it was on its behalf that on 14 June 1786, Knox was instructed to prepare to travel north to 'take the trouble of collecting the names of such persons as are willing to become subscribers to the fund of the proposed society for extending the fisheries…' (ibid.). There was a realisation that the Government alone was not going to pay for all the improvements in this remote region, and that local proprietors should be encouraged, for their own benefit, to contribute to this project, at least financially. Those who did so, Knox assured them, would find their names 'engraved upon every Highland rock, and be recorded with applauses [*sic*] to the end of time' (ibid.).

With his previous knowledge of Scotland, Knox was well equipped for the job, and he set off on 29 June 1786 'upon the hazardous and fatiguing enterprize. It was to travel, mostly on foot, from Oban in Argyleshire to Cape Wrath … from thence along the shore of the Pentland Firth to the North-East extremity at Dungsbay (Duncansby) Head' (Knox, 1787). He would then continue down the east coast, returning to Edinburgh via Aberdeen. Over the period of six months, he calculated that he had travelled 3,000 miles. Unlike that of Anderson, Knox's record of his tour contains as much detail about his travels as it does of his recommendations for improvements.

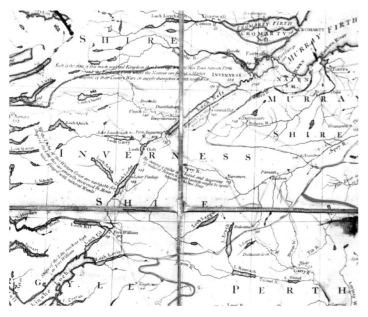

Knox had prepared a Commercial Map of Scotland as early as 1782. This detail shows his proposal for a canal linking Fort William to Inverness. Forty years later, the canal had been built, known as the Caledonian Canal.

Even as he crossed into Scotland from England, north of Carlisle 'a striking contrast is observable ... a picture of dreary solitude, of smoaky hovels, naked, ill cultivated fields, lean cattle, and a dejected people, without manufactures, trade, or shipping'. Heading north-east, he found the inhabitants of Edinburgh 'courteous, obliging, hospitable, and well inclined to the bottle, owing, it may be supposed, to their social dispositions and the excellence of their wines'. He then made his way back to the west. At Bunawe (Bonawe), he came across a smelting business run by an English company, and noted its influence. 'The verdant fields, and other agreeable appearances on this little spot, plainly indicate the residence of Englishmen'. But otherwise, he found 'nothing more than ruinous villages, exhibiting all the symptoms of decay, poverty, and distress' (Knox, 1787). On a more positive note, he observed, as Anderson had done, that a canal between Crinan and Lochgilphead would give shipping passing south down the west coast access to the Clyde, without the need to travel all the way round the Kintyre Peninsular. Similarly, he urged that a canal be built that would unite the three large lochs – Ness, Oich, and Lochy – that form the Great Glen. This would allow east/west shipping to avoid the treacherous waters of the north coast.

Like Bryce, Knox emphasised the dangers of the Pentland Firth, the narrow strip of water that lies between Orkney and the Scottish mainland. Even in the calmest weather there is 'a great swelling sea, with breakers during ebb-tide... In other parts the surges happen with the flood-tide. Ships endeavour to avoid the extraordinary convulsions of the water, but in some winds, they are forced directly among them ... Should a vessel be driven into the firth by the violence of a tempest, in the dark winter's night, [its] situation is dreadful beyond description' (Knox, 1787). For centuries, mariners had faced the difficult decision of whether to go north and round Orkney, a safer passage, or to cut through the treacherous Pentland Firth, thereby saving themselves 150 miles. The excellence of Mackenzie's charts had encouraged more of them to take the latter course, but still, 'it too often happens ... that by thus shortening the passage, they lose both ship and cargo, with their own lives, or the greatest part of them' (ibid.). Knox was told that in 1786, between April and October (the better season of the year), 11 vessels, mostly loaded, had been wrecked. The only beneficiaries were those who made their living from beachcombing.

As well as the tides and currents found in this stretch of water, two rocks called the Pentland Skerries had to be negotiated, as well as the Island of Stroma (see colour section, page iv). There had been a tradition in Scotland of burying the dead on small islands where there was less risk of wild animals interfering with the remains. On Stroma this was taken to a new level. The Kennedy family built a mausoleum there in 1677, which according to Knox, housed 'natural mummies, being the entire uncorrupted bodies of persons who had been dead above half a century; light, flexible in their limbs, of a dusky colour ... The coffins are laid on stools above ground ... And the rapid tides of the Pentland firth running by it, there is such a constant saltish air as hath thus converted these bodies into mummies' (1785). Bryce had remarked on his map that some have been 'lying there these threescore years and still entire.' Robert Forbes, the bishop of Ross and Caithness, reveals more. Murdoch Kennedy 'played such wretched tricks on the Body of his Father, for the Diversion of Strangers, as in time it broke to pieces, and the Head was the part that fell first off. He used to place Strangers at his Father's Feet, and by setting a Foot on one of his Father's, he made the body spring up speedily and salute them, which surprized them greatly. Then, after laying the body down again, he beat a march upon the Belly, which sounded equally loud with a Drum' (Forbes, Journals, quoted in Craven, 1883). By 1786, the *Hibernian Magazine* reported that the mummies had been trampled on and destroyed by sheep and cattle which had found their way into the mausoleum.

For those who are not sailors, it is hard to appreciate just how difficult navigating the northern oceans was, without the assistance of an engine. Knox found himself in alarmingly rough seas trying to get to Benbecula from Loch Bay on the west Coast of Skye, a distance of 25 miles. The water in the sheltered bay was calm enough, but when they reached the open sea they found themselves 'very unequally matched against the great swell of the open Atlantic'. The weather was deteriorating, and visibility rapidly decreasing but the master of the ship, Mr Macleod, was quite capable of steering by compass. Knox's fellow passenger, a stone mason, simply lay down upon his back, under the half deck, where he 'continued in that position, with his hands lifted up, and seemingly in fervent prayer, through the whole voyage'. Knox was more constructive. While Macleod watched the sails and shouted orders to his crew, 'my department was to observe the approach of every successive wave, and to give timely notice, that we might rise and fall with it, the only means of preserving us from being buried under

a watery mountain. Yet neither Macleod's skill, nor my vigilance, could prevent us from getting a brush now and then, en passant, which made Macleod stagger … and knocked myself down more than once' (Knox, 1787).

Perceiving that they were not going to make it to Benbecula, Knox suggested putting in at Lochmaddy in North Uist. They headed off in that direction at some speed, since the wind was in their favour, and eventually some high ground emerged out of the gloom ahead which they took to be the entrance to the Loch. Macleod immediately ordered the 'dram bottle' to be produced in celebration, of which they all partook, except the poor mason, who was presumably either too deep in prayer, or too ill to want any. However, the difficulties were not over. On entering the Loch, 'the wind blew with more violence than ever, and being

"Life on Board a Scotch Ship. The Cook, Captain, and Mait". An engraving by Paul Sandby, *1751.*

in this direction almost ahead, there was no possibility of getting to a place where we could land. The day was now advanced, and time was precious'. So, although they were yards from the shore, they had to tack about and search for some other port of safety. At one point it seemed as if they would have to go all the way to Stornoway, a distance of 50 miles, but in the end they were able to slip into Loch Rodel on Harris. The captain, who had often sailed round the Cape of Good Hope, 'declared that he would rather go there again, than come from Sky, in [sic] such a day, with an open boat' (Knox, 1787). Knox does not record what the mason thought.

However, on the mainland Highlanders preferred not to venture onto the sea. MacCulloch thought the Highlander 'neither a boatman nor a seaman, but a bear in a boat, a landsman at sea. He is naturally and essentially a farmer, and only a boatman by chance' (II, 1824). The 1799 *Statistical Account* agreed: 'Seamen properly speaking there are no such here, but if tugging an oar in a boisterous sea can be called an accomplishment of seamen, in this event all the tenants of the noble proprietors along the coast are seamen.' Observers were impressed by the strength of these men. Knox watched boats off Skye, returning from the herring grounds, rowing into the wind for 10 to 25 miles. 'They sing in chorus, observing a kind of time, with the movement of the oars … Those who have the bagpipe, use that instrument, which has a pleasing effect upon the water and makes these people forget their toils.' A romantic image, were it not such a struggle. If the men returned with herring, the effort was worthwhile, but if they had caught nothing, which was often the case, 'the disappointment to their half-starved families is easier to be conceived than expressed, and they [had] the same work to perform again, as soon as herrings [were] heard of, within the distance of fifty miles' (Knox, 1787).

Herrings were both a blessing and a curse to these regions. They could not be relied upon to visit the same waters each year, but they were plentiful when they did. James Robertson, the botanist, was told 'they come in such immense shoals that they wholly fill up the Bays and Lochs they enter into, so full that [the locals] can standing upon the rocks take them out of the water with their hands' (Robertson, 1994). Sometimes the nets on the boats broke under the weight of fish within them, leaving the sailors without their nets, and the sea 'infected with the stench of the dead fish' (Southey, 1972). At other times, the herrings would be driven into the bays by whales, which would then become grounded, giving the inhabitants a double catch.

The herrings brought in fishermen from all over Europe – Spain, France, and especially, Holland. For the northern population, it was a remarkable invasion in the summer months. The 1835 Penny Magazine reported that, for up to six weeks from July, 10,000 people descended on Wick in 1,200 boats. All sorts of tensions arose as a result of this invasion, not least with the Church which could not persuade the English fishermen to join them in their kirks on Sundays, and, worse still, could not persuade their European counterparts to desist from fishing on the Sabbath. 'The French and Dutch observe no distinction of days in fishing; and the minister has applied for a revenue-cutter to enforce the observance of the Sabbath-day, by breaking the nets of the delinquents'.

"Wick Harbour During the Herring Fishing". From The Illustrated London News, *1875.*

"Dutch Fishermen at Lerwick". A postcard sent from Wick in 1914. The sender notes 'We have seen lots of these men … as the Herring season is in full swing – the harbour crowded with Norwegian, Dutch, and other boats of every nationality.'

Herring gutters, Wick. A photograph by Valentine.

The Dutch were the most efficient of all the fishermen, preparing the herring on board as soon as it had been caught. The local boats preferred to bring the fish back to the shore, giving employment to 5,000 women who gutted and salted the catch at the docks. 'The herrings exported from Wick are chiefly intended for the subsistence of the poorer classes of the Scotch and Irish, and for the slaves in the West Indies, whither they are conveyed from Bristol' (*Penny Magazine*, 1835). The superior quality herring found a market on the Continent, and of course a better price.

There was no shortage of other fish in the Scottish waters. At a meal on Skye, Knox was offered 13 different types of fish, all of which were caught locally. But fishing was a dangerous business. Moreover, salt, which was taxed, was expensive, and the landowners forbade their tenants to fish the freshwater lochs, even those, says MacCulloch, that their proprietors 'never saw, and probably never [would]'. The result was that, in a land that suffered severe famine at times, fish did not form a large part of the poorer people's diet. 'Even in summer', continues MacCulloch, 'I have entered the Highland cottages hundreds of times, without finding fish, either in the act of eating or in possession [*sic*]' (III, 1824). Pennant observed that there are 'a few, a very few of the natives who possess a boat and nets'. Even such items as fish hooks could be in short supply, as he discovered on Rum. 'Their stock of fish-hooks was almost exhausted … It was not in our power to supply them. The ribbons and other trifles I had brought would have been insults to people in distress. I lamented that my money had been so uselessly laid out; for a few dozens of fish-hooks, or a few pecks of meal, would have made them happy' (Pennant, 1998).

Knox continued slowly up the west coast. At the barracks at Bernera, he found a 'groupe of mean huts, and the most miserable looking people that I had seen.' He was, however, 'entertained by the commanding officer, and his whole garrison. The former was an old corporal, and the latter was the old corporal's wife: the entertainment snuff and whiskey.' Clearly the garrison was not what it used to be, and Knox noticed the increased poverty evident since the withdrawal of the troops. A little further to the north, he gazed on the mountains of Skye from Raasay, finding them 'terrible to behold' (see colour section, page v), and mused on the height of the Himalayas which he reckoned to be as much as 8,000 feet. Between Applecross and Skye, he spotted a multitude of whales, the more distant ones visible by the water spouts they sent up into the air. All were devouring herrings by the hundred at a time, 'but many are thrown into the air with the water, and thus have a narrow escape'. There were grampuses [dolphins] too, which he watched as they tumbled and leapt out of the water.

Occasionally he would come across an efficient and successful business, such as the fish-curing house at Ardmore, on the south side of Loch Torridon, but on the whole inefficiency was the norm. Gairloch had a similar establishment, but when the owner 'opened the door

Crew of Whaling-station, Bunamhuinneddor. (Copyright.)

The Whaling Station on Harris. Run by Norwegians, this operated until 1920.

An engraving of the curing buildings at Ardmore, from Thomas Newte's Tour in England and Scotland 1791. *Newte ventured no further north than Inverness on his travels, but was interested to hear of the business at Loch Torridon.*

where … [the fish] lay, the smell was intolerable' (Knox, 1787). The fish had gone off before being salted.

Progress was difficult. Knox was in an area 'where no man, who cannot climb like a goat, and jump like a grasshopper, should attempt to travel'. He thought the mountains of Ross-shire 'convulsed into a chaos', but Sutherland was worse. 'The County of Sutherland is the most remote in Great Britain, and also the most rugged and least improvable.' Now, like Anderson before him, Knox found himself in October at this, 'the most difficult part of my enterprize … [when] the rivers might be swelled by rains, … [and] the swamps would be covered with water' (1787).

He was fortunate in finding a good guide, a half-pay officer named, like many in the area, Mackenzie, who offered to lead him to Durness, and further if necessary. Knox gladly accepted, though he had trouble keeping up with his 'Highland Trot'. 'I was continually in a sweat … while Mr Mackenzie … travelled with all the agility and ease for which his countrymen are remarkable.' They explored some of the region by boat, but crossed from Loch Inchard to Durness by land, which he thought a desert. 'Here are no trees, no houses, no people'. He saw huge stones and rocks lying on the hills which he imagined to have been transported with great effort and difficulty, probably 'to screen the persons who were on the watch to kill the wild boar, the deer, the fox, the eagle, and other animals' (Knox, 1787). The Ice Age, and the power of glaciation which shifted vast quantities of material, including large boulders, were subjects not yet understood at the end of the eighteenth century.

They were welcomed at Durness by a Mr Anderson, and at this point Knox's journal comes to an end. He had intended to write a second part detailing his journey along the

Travelling Map of Scotland. *This map, published about 1820, still relies on Bryce's description: 'The Moin, a deep Morass.' Once again, it shows a road that had yet to be built.*

north coast, but this was never published. However, in a footnote, he reveals that he crossed 'the Moan' (the Moine) between Eriboll and Tongue, with the assistance of two guides, and became another victim of its difficulties:

> [It was] a most fatiguing journey … being very tired, I went early to bed. The gentleman at whose house I lodged, was pleased to show me into the bed-room, which was full of smoke, owing to a newly kindled peat fire, and an unfavourable wind. When he went out, he left the door a little open, and advised me to let it remain so through the night. Instead of following his council, I shut the door, and took off the greenest peats, thinking by that means to get rid of the smoke. I went to bed and fell asleep in less than five minutes – Awoke about two o'clock in great agony, and scarcely able to breathe. Having a confused sense of my situation, and the cause of it, I attempted to rise and open the door, but in getting up I found much difficulty from want of strength. In going towards the door, I fell upon the carpet and lay there till the family, who had heard the noise, came into the room, where I was found cold, but with symptoms of life.

Once in a room free from smoke he recovered, and the next day felt only 'an oppression at my breast, with a headache' (Knox, 1787). But he had been within a few minutes of expiring.

Knox's journey had confirmed his worst fears: a country with impoverished villages, poor or non-existent lines of communications, and a people unmotivated and depressed. As for towns, Pennant before him had observed that 'there is not a town from Campbelton in the Firth of Clyde to Thurso, at the end of Cathness, a tract of above two hundred miles' (Pennant, 1998). Proceeding up the west coast and then along the north, it is in fact nearer 400 miles.

Anderson and Knox were very much in agreement over what needed to be done. The lines of communication should be improved – roads, bridges, and canals at Crinan and the Great Glen. Towns with facilities for passing shipping were required. Communities should be set up that were self-supporting, with a variety of crafts and trades people – blacksmiths, boat-builders, carpenters, as well as spinners and weavers – and fishing had to be properly organised and encouraged. They wanted to see an end to the feudal structure that prevailed, together with the short-term tenures of the rented land which deterred any attempt at improving the quality of the soil and its productiveness.

In the north, Knox had found 'a great body of people, and these the most virtuous of our Island, dragging out a wretched existence, perishing through want, or forced through wild despair to abandon their country, their kindred, and friends, and to embark moneyless and unknown, the indentured slaves to unremitting toil and drudgery in boundless desarts at a distance of 3,000 miles.' Those that were left, a people 'intelligent, hospitable, religious, inoffensive in their manners, submissive to superiors, temperate, frugal, grateful, obliging, honest, and faithful' (Knox, 1785) were to suffer further upheaval and ill-treatment over the next 50 years. However, the Government was poised to take positive steps to modernize the Highlands, and bring them into the fold. Whether this was from compassion, or a need for cannon-fodder is debatable.

"Stranded on the East Coast." A late 19th century photograph by George Washington Wilson. Said to be the Isabella, possibly from Downies, Aberdeenshire. It was wrecked on the 21st April, 1880, with the loss of six crew.

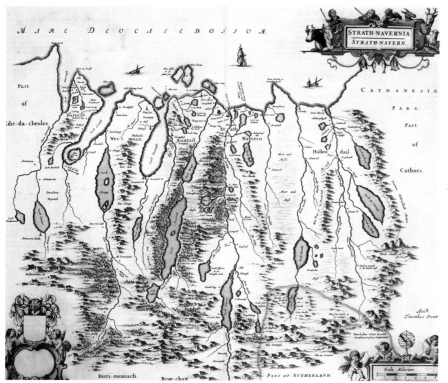

The map of Strathnaver, from Blaeu's 1654 Atlas, Theatrum Orbis Terrarum. Such details as 'A hole of 180 fathom deep' at the south end of Loch Eriboll, and 'Bin Staomny, whair is had iron oare' east of Loch Loyal were taken directly from Pont's survey.

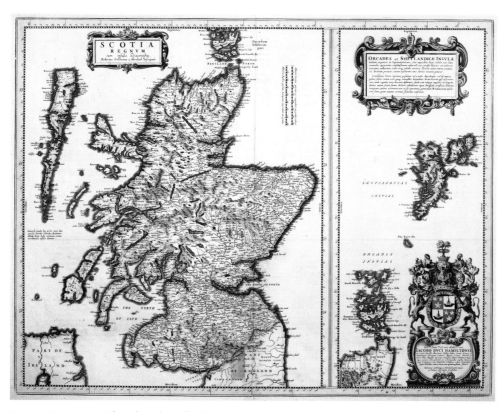

Scotia Regnum. *Blaeu's Atlas of 1654 carried this map of the entire country, by Robert Gordon of Straloch.*

I

Loch Maree, looking NW over the Cladh nan Sasunnach, or Englishman's Churchyard.

Left: The column from Bryce's "Epitome of the Solar System on a Large Scale", which stands in the Almondell Country Park. An Epitome is defined as a summary or miniature form.

Right: "Come Under My Plaidy". An original sketch with the monogram 'MOG', illustrating a rather sinister poem by Hector MacNeil. Both the older man and the younger suitor are dressed in full Highland outfit.

II

"The Antiquarian Society". An engraving by George Cruikshank, published in 1812. It may depict General Roy, standing at the far right of the image.

"The Scotch Cottage of Glenburnia". An engraving by Isaac Cruikshank, commenting on the state of Scottish Houses. 'Mistress McCarty, why do you not make your daughters assist you?' asks the visitor, '…it's no often they will be Fashed [bothered]' the lady of the house replies.

Left: "Girl with Snood". An anonymous 19th century drawing.
The snood [ribbon] denotes that she is unmarried.

Right: Eagles soaring at Braeriach in the Cairngorms. A detail from a painting
by Samuel J. Barnes. The artist's work was much admired by Queen Victoria.

"John O' Groats, Caithness". An engraving by William Daniell, 1821.
The lighthouse seen in the distance marks the Pentland skerries, while
the cliffs on the far left horizon are possibly those of Stroma.

"From the Isle of Rasay Looking Westward." An engraving by William Daniell, 1819.

An original sketch by Arthur Perigal (c.1880). The fishermen are launching the boat from the shore. Knox noted the difficulties of such a manoeuvre, and recommended the building and improvement of harbours throughout the area.

Above: "Posting in Scotland". A Georgian caricature print by C. Loraine Smith (pseudonym of James Gillray), 1805.

Left: Telford, depicted in the frieze by William Hole at the National Portrait Gallery in Edinburgh. Behind him can be seen the geologist James Hutton (image courtesy of the National Galleries of Scotland).

Above: In 1792, J. Wallis published A
New Geographical Game exhibiting
a Complete Tour through Scotland
and the Western Isles. *It was a simple
game which involved moving counters
from number to number on a map.
There were 100 numbers in all. North of
Fort William and Inverness there were
eight numbers along the East Coast
– Cromarty and Dornoch, amoungst
others. In the North-West mainland
region, there is just one: number 16 –
Assynt – which, as it states in the key
'exhibits an assemblage of shattered
mountains, as it were heaped on each
other, and seemingly involved in a
tremendous manner.'*

*Right: Murchison, on the frieze by
William Hole, at the National Portrait
Gallery in Edinburgh (image courtesy of
the National Galleries of Scotland).*

*Nairn Harbour, by Arthur Perigal (c.1882). The harbour
at Nairn was one of those improved by Telford.*

*"Ilan-dreoch Glenbeg, Inverness-shire." William Daniell's print showing the
stretch of water over which he was carried by a 'mountain-nymph.'*

*"Bonar Bridge". A print by William Daniell, 1821. It was
described by a local as 'something like a spider's web in the air…'.*

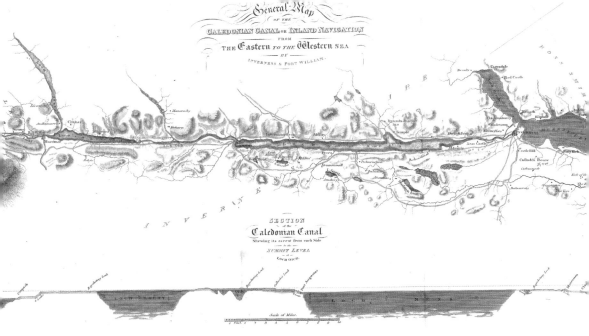

General Map of the Caledonian Canal. *Engraved by W.&D. Lizars, published by Constable & Co., Edinburgh, 1817.*

An unsigned original watercolour of a Highlander heading off to work. He carries heavy tools, such as a pickaxe and spade, which would not normally be used for agriculture in these parts. Perhaps a roadbuilder?

Above: Reminders of the road building projects can still be found in the north west. This inscription was added to a well on the road south of Durness: 'As a mark of respect to the inhabitants of Durness and Eddrachillis for their hospitality while projecting this road, this inscription is placed over this well by their humble servant, Peter Lawson, surveyor.'

Below: The remains of Moine House, with Ben Loyal in the distance.

A pair of aquatints by and after J. Clarke, 1828.

"Doctor Prosody Attacked by Soland Fowl in the Orkneys".
An engraving by W. Read, 1821.

"Ben Nevis, from the steamer. 8 a.m., August 1858." A sketch by an anonymous tourist.

"Doctor Prosody Proves the Inconvenience of a Timid Companion at Staffa." An engraving by W. Reid, 1821.

Doctor Prosody Meets a Highland Wedding on the Caledonian Canal. An engraving by W. Read, 1821.

"Loch Awe – between Dalmally & Inverary." A sketch monogrammed "FJ" and dated Oct. 25th 1811. Faujas found the road 'melancholy and painful'.

Left: "*Highland Chiefs, Dressed in the Stewart and the Gordon Tartans.*" *The frontispiece from the first volume of James Logan's* Scottish Gael.

Below: Carte Géologique de l'Écosse, *the first geological map of Scotland, from A. Boué's* Essai Géologique sur l'Écosse, *1820.*

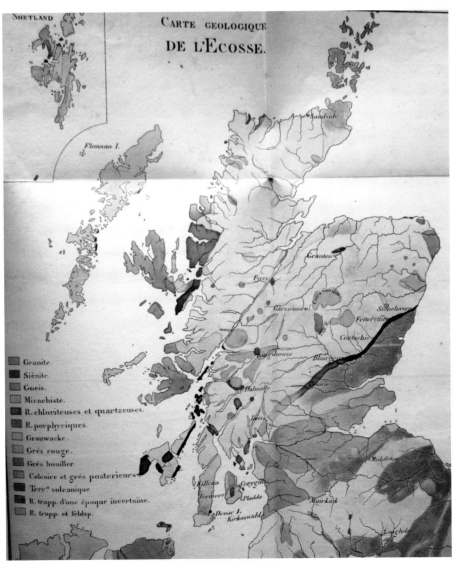

Left: "Rispond, Durness." An engraving by William Daniell, 1820, looking across Loch Eriboll to Whiten Head.

Lower left: "Castle Ellen-Donan." An engraving by William Daniell, 1818. Eilean Donan castle was in ruins until the early 20th century, when it underwent full restoration. It is now one of the most photographed sights in the Northern Highlands.

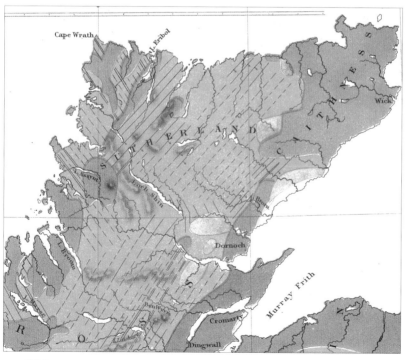

Detail from a Map of Scotland, 1852, by Daniel Sharpe. The lines represent the direction of the bands within the rocks: the fact that those from Cape Wrath go NW to SE, while those to the east of Loch Eriboll are NE to SW suggests two different systems.

Frontispiece from Siluria, *the new edition of Murchison's flagship volume (1859 edition). All the lighter-coloured rocks (labelled with a small 'c') were claimed as Silurian by Murchison.*

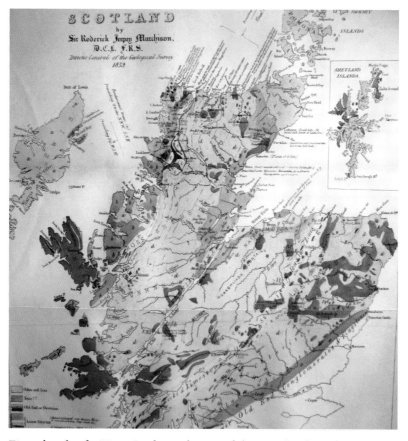

First sketch of a New Geological Map of the North of Scotland, *by Sir Roderick Impey Murchison. From the* Journal of the Geological Society of London, *1859. This was the first map to show Murchison's view of the geology of the Highlands.*

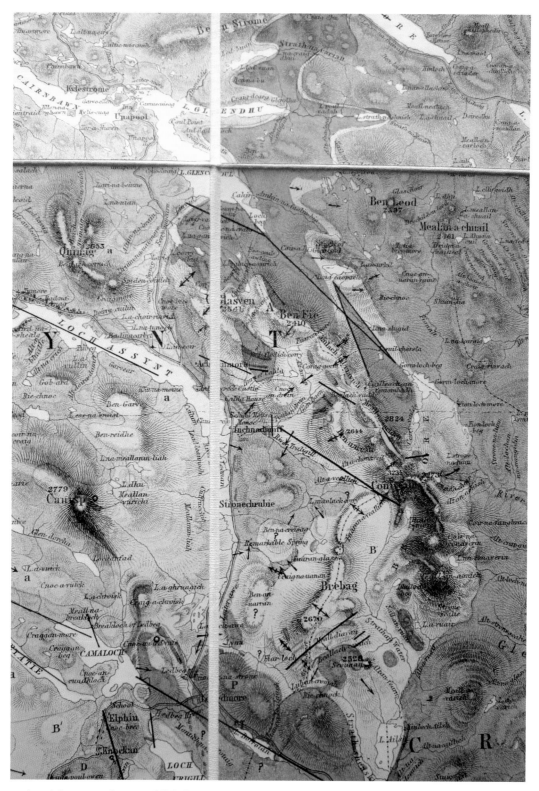

A detail from Matthew Heddle's fine Geological and Mineralogical Map of Sutherland, *published in 1882. Areas in which the Logan Rock was to be found are marked in a bright green.*

CHAPTER FIVE

THE COLOSSUS OF ROADS

The works conducted by the Parliamentary Commissioners from the year 1803 have done more to advance the civilisation of the highlands than all the other attempts of Government for that purpose in the course of the preceding century.

(*Penny Magazine*, 1833)

The north of Scotland was not the only part of Britain which had to contend with poor roads. The Sussex Weald was a notoriously muddy area in winter. The Duke of Denmark, in 1703, took 14 hours to cover the 40 miles from Windsor to Petworth in his coach, the last nine taking six hours. Travellers on the main road south from Glasgow in 1739, according to the *Penny Magazine*, found 'a narrow path raised in the middle of an unmade soft road, into which latter they had to descend whenever they met one of the gangs of packhorses carrying goods, the raised causeway not being broad enough to allow the two parties to pass each other' (*Penny Magazine*, June 1833). Further west, on Andrew Armstrong's map of Ayrshire (1775), a note advises that 'at the Nick of Darlae and half a Mile West, the road leads on the Side of a very steep Hill: its [*sic*] not above two feet broad, and if you stumble you must fall almost Perpendicular six or seven Hundred Feet.' Few of the roads were suitable for wheeled vehicles, and Robertson's *Rural Recollections* record that in 1763 there were only three stagecoaches in the whole of Scotland, all of them based in Edinburgh. Two of them formed a regular service down to Leith, a distance of three miles, whilst the third went south to London once a month, the journey taking 12 to 14 days.

However, at least there were roads in these Lowland districts: in the far north there were scarcely any. Throughout the eighteenth century the main highway into Sutherland was the *only* highway in that county. It clung to the east coast for 30 miles from Meikle Ferry to Helmsdale, providing a passage that 'never failed to inspire individuals not accustomed to such passes, with great dread' (*New Statistical Account*, 1845). Robert Forbes (1708–1775), the episcopal bishop of Ross and Caithness, travelled along this road on a journey he made to Thurso in 1762, and he provided a more detailed description. Having left his wife in

Inverness as he knew just how poor the roads would be, and having forded the river at Helmsdale, he was then faced with Ord Hill, and proudly announced that he 'rode up every inch of it, a thing rarely done by any persons ... Its steepness, and being all along on the very brink of a precipice are the only difficulties'. It was so steep 'that no machine can be drawn up it by any cattle [*sic*] whatsoever, unless it be empty; and even then, there must be some sturdy fellows at the back of it pushing it forward to assist the horses; for if they are allowed to make the least stop, backward they must tumble by the very declivity of the place' (Forbes, *Journals*, quoted in Craven, 1883).

Rather than follow this road all the way along the coast, the bishop chose to cut across on the direct route to Thurso, known as the 'Causeway-mire', for which he needed a guide. 'Why it is called such a name I could not conceive', he wrote, 'as the smallest vestige of a causeway we could not discover in the whole'. He found he had to keep the horses going at all costs, 'for if they make the least halt, or too leisurely a step, down they must sink'. He was informed later that 'a causeway had been there of old, but it had sunk down out of sight by the ruins of time' (Forbes, *Journals*, quoted in Craven, 1883).

The first road atlas of Scotland, published by Taylor and Skinner in 1776, shows both the coastal road to Thurso, and the Causeway-mire, branching off to the left. There are no roads at all in the north-west corner of Scotland, an area that comprises much of Ross-shire and Sutherland. Wade's roads had given a flavour of the advantage such lines of communication could lend to these remote parts, but as we have seen, they went no further north than Inverness. What is more, in 1799, the Superintendent of Military Roads in the Highlands, Colonel Robert Anstruther, reported to the Government that these roads were in an appalling state of repair.

Others were lobbying hard for action, not least George Dempster of Dunnichen, 'Honest George' as he was known, 'one of the most conscientious men who ever sat in Parliament' (Wraxall, 1836). He became chairman of the Highland Society in 1789, and, in his later years, left Parliament to devote much of his time, and his fortune, to improving the lot of the rural population on his estates at Skibo and Dunnichen. The ruins of a mill that still stand in Spinningdale are testament to his efforts, though that particular project was a failure, and it burnt down in 1806. He was more successful in obtaining a promise from William Pitt that the Government would look into the situation regarding Highland roads, and in 1801 the Joint Secretary to the Treasury, Nicholas Vansittart, contacted Thomas Telford, inviting him to proceed with a survey to assess the state of the Highlands generally at that time.

Telford was born in Dumfriesshire in 1757. Largely self-taught, he began his working life as a stone mason. One of his first projects was the headstone for the grave of his father who had died soon after Thomas's birth. Telford was clearly a very likeable person: 'of the most genial disposition' wrote one of his draughtsmen after his death in 1834, 'and a delightful companion, his laugh was the heartiest I ever heard; and it was a pleasure to be in his society' (Turnbull, 1893). In 1782 he made his way to London, and caught the eye of such distinguished employers as Robert Adam and Sir William Chambers, both architects at the top of their fields. By 1801, he had made his reputation on his own with such projects as the Ellesmere and Shrewsbury Canals, and the restoration of the bridges at Bewdley and Tenbury, both of which had been swept away in the floods of 1795.

Telford was greatly motivated by this Government initiative in 1801, and had set off on the survey even before he had received his full instructions, such was his desire to serve his homeland. He confessed, 'Never when awake – and perhaps not always when asleep – have my Scotch Surveys been absent'. His tour was blessed with exceptional weather, even into the autumn. 'I passed along the Western and Central Highlands, from thence to the extremity of the Island, and returned along the Eastern Coast to Edinburgh, and scarcely saw a Cloud upon the Mountain's top'. On his arrival back at the capital, he announced 'My plans are completed … may they be productive of good and that good benefit Scotland' (Letter, 1802, quoted in Turnbull, 1893).

The Government requested a second survey, which was submitted in 1803, shortly after which two bodies were set up: The Highlands Roads and Bridges Commission, whose remit also included improving harbours and towns, and The Caledonian Canal Commission. The project was set in motion, and work began in 1804. Over the next 17 years, 892 miles of road, 1,300 bridges, as well as harbours, churches, and a hugely ambitious canal would be constructed, all in some of the most difficult terrain in the British Isles.

Central to the plan was the completion of a road all the way to the north coast via Inverness – what was to become the northern end of the Great North Road, now the A9. Spur roads leading off to the west would open up the remotest parts of the country. The Commission calculated that a proper route from Skye to Killin in Perthshire would save proprietors £12,000 a year in 'Droving Expenses, and in avoiding unnecessary deterioration of the Animals driven' (5th Commissioners' Report, 1811). Such inducements persuaded the landowners and counties to give generously towards the costs, while the Government provided a half share of £267,000. Most importantly, the Commission made sure that the scheme was properly endowed; it secured an annual grant of £5,000 a year from Parliament for the repair and upkeep of the roads.

Telford chose as his assistant John Duncombe, with whom he had worked at Ellesmere on the canal. It turned out to be an unfortunate choice, though it is not exactly clear what happened. Not only did Duncombe's health deteriorate, but by 1809 he was incarcerated in Inverness prison, where he died. Telford's reaction was unusually ungenerous – he must have suffered much irritation over Duncombe's behaviour. 'His dying', he said, 'will not be a matter of regret, but in a Jail at Inverness is shocking' (letter to Rickman, quoted in Turnbull, 1893). Duncombe was succeeded by John Mitchell, whose service was exemplary. Robert Southey, the poet, extolled him as a man of 'inflexible integrity [with] a fearless temper and an indefatigable frame.' He needed the indefatigable frame: he is said to have covered over 8,000 miles in a year in connection with his duties. Southey had a lift in his gig at one point. 'Had the distance been a few miles farther I believe neither my poor pantaloons, nor my poorer flesh, nor the solid bones beneath, could have withstood the infernal jolting of this vehicle, tho' upon roads as smooth as a bowling green. As for "M" [Mitchell], he is so case-hardened that if his horse's hide and his own were tanned, it may be doubted which would make the thickest and toughest leather' (Southey, 1929).

James Mitchell died in 1825. His son Joseph stepped into his father's shoes and his two-volume *Reminiscences* provide a vital record of the project. Further sources of information are the various reports issued by the Commission during the works.

Telford divided the Highlands into six divisions, each with its own superintendent. Contracts for the various roads were then put out to tender. The Commission quickly learned 'how much more important is the personal character of a Contractor than any cheapness of a proposal' (7th Commissioners' Report, 1815). At Loch-na-Gaul (Loch nan Ceall) Mr Readie was sacked because the Commission considered 'his progress to have been so dilatory' (5th Commissioners' Report, 1811). At Livingstone, the contractor deserted, aware that he had grossly underestimated the costs involved, and the same thing happened on Skye two years later. In both cases the Report commends the local proprietors for taking over the extra cost involved in finishing the work, deeply regretting that the Commission lacked the power and the means to remunerate their loss. On the other hand, the contractors Messrs Simpson were commended for their reliable punctuality on the difficult Loch Ness section, and at Skibo, the work of Messrs Gilchrist and Peacock was seen to be 'a laudable example of what may be done in Highland Roadmaking' (5th Commissioners' Report, 1811). They were later rewarded with the contract for the Tongue road.

There were, inevitably, a number of unexpected problems. Disturbances occurred at Corpach when the wages failed to turn up. The local supervisor, John Telford (no relation), was agitated in the extreme. 'If the men are not all settled with Monday night at farthest, I dread the consequences' (Rolt, 1958). Thomas took a more philosophical approach. 'Misunderstandings and interruptions must be expected in works of this kind amongst a people just emerging from barbarism' (ibid.). At Glengarry, work came to an abrupt halt following a fatal accident which 'terminated the life of Mr Dick, the acting Contractor on this Road by his falling down a precipice on his way from Loch Hourn' (5th Commissioners' Report, 1811). Often, materials had to be carried over long distances, such as on the Tongue stretch where mortar was transported for 20 miles by pack horse. Sometimes, surprisingly, suitable stones were not available in the vicinity; this was the case on Skye, for example, and also at Lovat bridge, where Southey noted that 'the foundations were expensive, and the stone was not at hand' (Southey, 1929).

Bridges formed a critical part of the whole project; they often presented major engineering challenges, but they were of paramount importance. Highland river crossings were a notorious problem, the ferries inefficient and the fords unreliable – what at one moment was a shallow trickle could become an impassable torrent within a couple of hours, sometimes caused by heavy rain several miles away. Osgood Mackenzie knew the innkeeper at Kinlochewe, who

Easter Fearn Bridge on the B7196, the Struie road south of Bonar Bridge. One of many bridges designed by Telford and his team which required intricate engineering. In this case 'The bridge … is remarkable for its tall approach walls of coursed large round rubble stones on a splayed base to provide stability' (Paxton & Shipway). It consists of a 40 foot span single arch which stands 50 feet above the Easter Fearn Burn.

had lost 'both her husbands … [who had] been drowned in trying to get people across this wild river on horseback when it was in flood' (Mackenzie, 1949). Donald Sage describes in his memoirs how he nearly drowned whilst crossing the river Carron. The previous day there had been 'not the slightest difficulty in getting over', but there had been heavy rain overnight. Unaware of the danger, he entered the ford, 'and had not ridden ten yards into the stream when [his] horse suddenly lost its footing, and [they] were both at once swept down by the strength and rapidity of the current'. On that occasion they managed to recover and struggle back to the shore, and a local who was passing guided him to a safer spot. Sage's father was completely fearless. When trying to get to church with his kirk-officer, and coming upon a swollen river where men on the opposite bank advised him not to cross, he seized his companion by the collar, and 'deliberately walked into the stream with him; then, taking a diagonal course against it, amid the roaring of the torrent, and the warning and almost despairing shouts of the men on the other side, he pushed on, and in less than ten minutes, placed himself and his gillie safely on the opposite bank' (Sage, 1899).

Osgood remembers that 'sometimes the children were carried over by men on stilts, which was thought great fun by them' (Mackenzie, 1949), and James Hall noticed at Kirkmichael 'a stout handsome girl tucking up her petticoats, and walking on stilts, perhaps four or five feet high, carrying her mother across the river to church' (Hall, 1807). William Daniell, who undertook a tour round the coast of Britain between 1814 and 1825, found himself at Glenbeg (see colour section, page viii), wondering how to cross some fast-flowing water, when

a robust highland female presented herself; and it was on the shoulders of this mountain-nymph that the passage was effected. Though the stream was so strong that an ordinary traveller would have lost his footing, for even a horse that was

Above: Patience was needed at the point of embarkation. Here, tourists jostle with sheep, seeking priority for boarding at Strachus – probably Creggans, Argyll.

Upper left: Even Royalty were in danger of getting their feet wet. A portrait by Allan Stewart dated 1901 of "Her Majesty and the Prince Consort Fording the Garry."

Lower left: Crossing a River to go to Church. An engraving from Hall's Travels in Scotland… 1807. *Hall's journey is described in Chapter 6.*

crossing at the time was observed to make considerable lee-way, she forded firmly and securely, without deviating from the direct line, and reached the other side in safety. She declined any pecuniary recompense, observing that this was an act of civility … and the only acknowledgement that she would accept was a small gratuity for her children. It was amusing to see how suddenly her scruples vanished, at the idea of obtaining a trifling gift *to please the bairns* (Daniell, 1978).

There might be a ferry at some of these crossing points, but access was rarely easy. Thomas Garnett, travelling with a companion through the Highlands after the death of his wife in childbirth, had to be carried out to the boat by a boy aged, perhaps, 16. He then watched as a younger boy, maybe 14-years old, 'got both our saddle bags, which were very well filled and heavy, and was taking them, as I imagined, for his share of the burthen; but to our mutual astonishment, he, thus loaded, made towards my friend, and mounting him on his back, ran with him and bags to the boat with much agility' (Garnett, 1800).

Alexander Brodie wrote in 1748 when he was considering visiting Balnagowan, 'The Ferrys [sic] at this time of year frighten me' (Macgill, 1909). He wasn't the only one to have a low opinion of them. 'The boats appropriated for these ferries are generally old and rotten', writes John Knox. 'Neither can horses get in or out without the risque [sic] of being lamed, owing to the great height of the banks, and their want of slips' (Knox, 1785). John MacCulloch found himself on a boat where 'all the arrangements were of the usual fashion; no floor, no rudder, no seat aft, oars patched and spliced and nailed, no rowlocks, a mast without stay, bolt or haulyards'. It eventually landed him 'like a mutinous Buccaneer, on some rocks; which proved, in the end, to be Sky[e]' (MacCulloch, I, 1824).

Patience was needed when waiting for the ferryman to answer your call – what MacCulloch termed *Patientia Ferryboatica*. On a sportsman's tour of Sutherland in the 1840s, Charles St John had trouble attracting the attention of the ferryman at Tongue. 'We after some delay were answered by the arrival of a small sailing-boat, far too narrow to take over either horse or carriage. On making inquiry of the ferryman, he told us that *the* large boat was at present out of the water and under repair, but that if we could wait ten days or a fortnight it might be ready. This did not sound promising…' (St John, 1884). They solved the problem by leading the horse over a fordable part of the sands at low water.

Dorothy Wordsworth, William's sister, made a tour of the southern Highlands in 1803. She devotes a long section in her account to her experience at a ferry at Loch Etive, which sums up the frustrations and problems at such places for travellers in the early 1800s. First, they had to wait 'a considerable time, though the water was not wide, and our call was heard immediately. The boatmen moved with surly tardiness, as if glad to make us know that they were our masters'. Their treatment of the horse was immediately rough and unhelpful, and the beast was already terrified before they had started pushing and shoving it over the ridge of the boat, there being no convenient opening:

A blackguard-looking fellow, blind of one eye, which I could not but think had been put out in some strife or other, held him by force like a horse-breaker, while the poor creature fretted and stamped with his feet against the bare boards … when we were just far enough from the shore to have been all drowned he became furious, and, plunging desperately, his hind-legs were in the water, then, recovering himself, he beat with such force against the boat-side that we were afraid he should send his feet through. All the time the men were swearing terrible oaths, and cursing the poor beast, re-doubling their curses when we reached the landing place, and whipping him ashore in brutal triumph … We had only room for half a heartful of joy when we set foot on dry land, for another ferry was to be crossed five miles further (Wordsworth, 1981).

When they reached the next crossing point, the ferryman was nowhere to be seen. Dorothy, who was 'faint and cold', went into the ferry house to sit by the fire. 'Though very much needing refreshment, I had not heart to eat anything there – the house was so dirty, and there were so many wretchedly dirty women and children.' What she found worse was 'a most disgusting combination of laziness and coarseness in the countenances and manners of the women, though two of them were very handsome.' At one point, while they waited for the ferry, the horse, which was still in a state of fright, bolted off with the cart. Luckily, he was stopped before any serious damage was done. The ferryman eventually appeared, and they resolved to swim the horse over behind the ferry. This being an unusual sight, the entire household turned out to watch as the horse 'strove to push himself under the boat'. However, they made it to the other side, where the ferryman demanded as part-payment 'eighteen-pennyworth of whisky' of which, Dorothy noted disapprovingly, the women 'partook as freely as the men'. Following these adventures, the horse became something of a liability; it panicked whenever they came to water, and when it took fright in Glen Coe, it threw brother and sister Wordsworth: 'the horse dragged the car after him, he going backwards down the bank of the loch, and it was turned over, half in the water, the horse lying on his back, struggling in the harness, a frightful sight!' (Wordsworth, 1981). To their relief, the horse was unharmed, and they were able to patch up the carriage and proceed.

The Highland Commission had every reason to be aware of just how dangerous the ferries were. On 16 August 1809, James Mitchell was rushing to catch the Meikle ferry, an important service for those travelling north over the Dornoch Firth. To his dismay, he saw it pulling away, ignoring his shouts as he galloped down the hill. When he reached the shore, his disappointment turned to horror: there was no sign of the boat. It had capsized and sunk almost immediately, drowning all but three. The sea had been calm, but the boat was hopelessly overladen. The disaster 'created a deep sensation all over the country' (Sage, 1899). Robert Southey met a local on the shore, who had vowed never again to go on a ferry following this accident, which effectively cut him off from the south, until a bridge was built at Bonar by Telford. What a bridge it was too! This same local thought it 'something like a spider's web in the air … it is the finest thing that ever was made by God or man!' (Southey, 1929) (see colour section, page viii).

The arrival and assembly of this bridge over the Kyle of Sutherland must have seemed to those watching more like an Act of God than man. It was pre-cast in Wales by William Hazledine, the *Inverness Journal* noting on 7 August 1812, that it 'is now reared for public inspection … in front of his foundry at Plaskynaston, where it forms a new object of attraction and wonder.' The bridge was then shipped by canal to Chester, thence by sea to the Dornoch Firth, arriving 'safe in the beginning of September' 1812 (6th Commissioners' Report, 1813). It was predicted it would remain 'a monument of public munificence, and of judicious enterprize in the Heritors of Sutherland' (ibid.), and all were encouraged when, soon after its construction, it survived a combination of ice and felled logs which created 'a formidable instrument of destruction … the crash of the Timber was heard at considerable distance; but the Bridge stood firm' (7th Commissioners' Report, 1815). Alas, Telford's bridge

Little had changed regarding small ferry crossings even by 1900. The American visitor to Sutherland, Emily Tuckerman, took these two photographs. She notes on the back 'This is the way we went by the ferry from Tongue to Melness, Scotland.'

The ferry at Dornie, Loch Long. An anonymous photographer (c.1880).

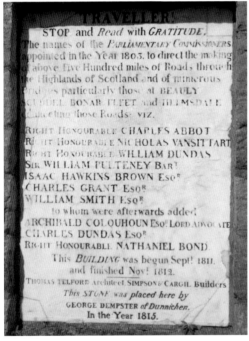

Above: "The Kyle [of Sutherland] & Bonar Bridge April 24, 1891. A grey day". A rare view looking towards the bridge, painted one year before it was swept away in floods. Artist unknown.

Left: The original plaque which now stands at the north-eastern end of Bonar Bridge.

eventually succumbed to a similar battering in the floods of 1892. The bridge that now crosses at this point is the third to have been built there, but the original plaque erected by 'Honest George' Dempster can still be seen at the north-eastern end of the bridge. 'Traveller, Stop and Read with gratitude the Names of the Parliamentary Commissioners appointed in the year 1803 to direct the Making of above Five Hundred Miles of Roads thro' the Highlands of Scotland…' Southey complained that it is 'full of errors both as to the Commissioners,

Left: "Saw Mills at Kilmonarck on the Beaulie River". A Georgian sketch by an unknown artist.

Lower left: "The Bridge at Potarch". A 1912 postcard. The original bridge was swept away in 1812.

Lower right: "Craigellachie Bridge". A photograph published by Valentine, (c.1885).

and the extent of the roads' (Southey, 1972) and he was surprised that it was attached to the toll house, not the bridge. The existence of a toll house suggests that tolls were collected, but the house was never occupied. There was not enough traffic to support a toll keeper, and travellers passed through free of charge. At Lovat Bridge, an Act had been passed allowing tolls to be charged, but they had never got around to building a toll house; at Helmsdale, 'there is such a house, and it has been let; but the man who rents it never demands toll: the house and the privilege of selling whiskey are considered by him as well worth the rent he pays, and he gives up the toll as not worth the trouble of collecting it' (ibid.). Such was the Highland attitude to road charges.

The new bridges proved to be vulnerable structures, not least because timber merchants were inclined to use rivers in flood to transport large logs which they would leave on the banks, hoping the waters would do the job for them. The bridge at Potarch was swept away in 1812. Floating debris was again the cause, but in this case the culprits agreed to pay some compensation and the crossing was back in action by 1813. Other bridges, such as that at Contin, were either poorly constructed or the design was inadequate to withstand the pressures that a Scottish winter could present. The bridge over the Spean had to be redesigned: Telford admitted that he had learned the hard way that the required span could only be gauged by studying the rivers during the rainy season. On the other hand, the Commission could be proud of designs like that at Craigellachie, another iron structure using the latest technology, cast again by Hazledine in Denbighshire. The huge sections of metal were transported by sea to Speymouth, thence by wagon to Craigellachie. The residents must have been amazed when such traffic trundled down their usually silent tracks, and maybe too by the design of the bridge with its mock-medieval castellated towers at each end and single cast iron span arching over the Spey, where it can still be seen today.

Another novel solution to the problem presented by water crossings was the Mound at Fleet, a huge causeway embankment nearly 1,000 yards in length and the base 60 yards wide at the east end. The water which passed through was controlled by sluice valves fitted under a bridge with five arches. A similar causeway was suggested for the Kyle of Tongue, which would have saved Charles St John his two-week wait for the ferry, but this scheme was not put into effect until 1971.

The most ambitious project was undoubtedly the construction of the Caledonian Canal. Work further south on the Crinan Canal had begun as early as 1794, under the supervision of John Rennie, another Scottish engineer of some note. The construction challenges there were of a far simpler order, but the work quickly ran into various problems – the weather, the workforce, the finances – and it was only completed in 1801, two years behind schedule. Even then there were various faults, which were passed to the Commission, and were eventually corrected by Telford in 1816. The canal's finest hour was in 1847, when Queen Victoria enjoyed a trip down it in some style, after which the steamship companies exploited its attraction as a 'Royal Route'.

The Caledonian Canal was altogether a different proposition, requiring in total 22 miles of man-made waterway linking the three lochs (Lochy, Oich, and Ness) which form the Great Glen. Over the entire 60 miles in length, the water had to be raised to a height of 100 feet, with the steepest ascent needed two miles from Corpach. Here, it had to be raised 72 feet, and Telford's solution was 'Neptune's Staircase', a set of eight locks at Banavie, one of the longest and most impressive stretches of masonry of any canal in the country. A minimum depth of 20 feet was required along the length of the canal, which meant a new piece of machinery was needed, the dredger. The *Inverness Journal* on 21 November 1817 reported that it was damaged when it came across some huge oak trees 'in seven feet of water and buried under a depth of ten feet of gravel.' Eventually it dragged three huge trees to the surface, the wood apparently 'perfectly fresh and sound.' The biggest problem, though, was at the Inverness end, where, to maintain this 20-foot depth out into the sea, the canal had to be extended by 200 yards, through a bed of mud 60 feet deep. Before work began, the cost of the canal had been estimated at under £500,000, but by the time the canal opened in 1822,

"Neptune's Staircase." The set of eight locks at Banavie.
A photograph published by Valentine (c.1880).

it had risen to over £1,000,000. Alas, it was never a great commercial success. James Hogg, on his tour in 1803, pointed out that the traffic passing from the east coast down to the west would be sailing into the prevailing wind – not ideal if you were relying only on sail. But as an example of early nineteenth-century endeavour, it was a triumph. Moreover, it gave employment to nearly 3,000 Highlanders over a 20-year period.

All this work is well documented in Joseph Mitchell's *Reminiscences*. The masons came from Nairn or Morayshire, while the labourers were Highlanders: two different types. Most of the masons led a sober life, with no spirits or beer, and they were able therefore to save a sizeable part of their 21-shilling wage. The Highlanders were not so provident. On pay day, once a month, they would head straight for Fort William where 'some drank continuously from Saturday till Monday or Tuesday night, without food and without sleep, beyond dozing on the table' (Mitchell, 1971).

The masons' accommodation was simple; so too were their meals. Work began at six in the morning, with breakfast at nine – 'porridge and milk and thick oaten bannocks about half-an-inch or three-fourths of an inch thick'. Lunch was at two o'clock, a repeat of the breakfast menu, but supper offered seasonal produce, maybe fresh herrings from Loch Hourn, and new potatoes. Again, the Highlanders differed: they survived on brose, 'meal in a bowl, a little salt, and hot water mixed into a mess' (Mitchell, 1971).

To capture the real spirit of this monumental work, we must turn to the account left by Robert Southey. His tour of Scotland took place in 1819, by which time he was a well-known figure given that he had been Poet Laureate since 1813. He met Telford through a mutual friend, the civil servant and statistician John Rickman, and was at once drawn to the engineer. 'There is so much intelligence in his countenance, so much frankness, kindness and hilarity about him, flowing from the never-failing well-spring of a happy nature, that I was upon cordial terms with him in five minutes' (Southey, 1972). Southey's disinterested enthusiasm and intelligent powers of observation bring alive this astonishingly complicated project.

Photograph of a portrait of Robert Southey, published by Pettit of Keswick.

At Fort Augustus, he found 'such an extent of masonry, upon such a scale, I have never before beheld ... it was a most impressive and rememberable [*sic*] scene. Men, horses, and machines at work; digging, walling, and puddling going on, men wheeling barrows, horses drawing stones along the railways. The great steam engine was at rest, having done its work ... the dredging machine was in action ... its chimney poured forth volumes of black smoke.' He descended into the dredger 'and saw the works; but I did not remain long below in a place where the temperature was higher than that of a hot house, and where the

"Locks on the Caledonian Canal at Fort Augustus". A photograph by George Washington Wilson.

*"The Caledonian Canal at Muirton." The termination of the
Canal at Inverness. Photographer Valentine (c.1880).*

machinery was moving up and down with tremendous force, some of it in boiling water.' The sight of the locks at the Inverness end of the canal was so exhilarating for his party that they stood up and gave three cheers as they crossed the drawbridge. He described being in the lock 'as hint for a Giant's dungeon ... such walls of perpendicular masonry ... such tremendous gates.' The sluices that release water 'brought the mightiest of the Swiss waterfalls to my recollection', but it was not just the noise and magnitude of the operations that impressed him. He wondered at the nonchalance of the engineers. On seeing some large chunks of iron lying around, 'it seemed curious to hear Mr Telford talk of the propriety of weighing these enormous pieces (several of which were four tons in weight) and to hear Cargill reply that it was easily done.' He was surprised that in all the diggings for the canal, some of which were 'cut forty feet below the natural surface of the ground ... nothing has been found except a silver chain, which was near three graves in the line of the Canal, near Inverness' (Southey, 1972). Southey may not have been entirely correct on this matter, for the *Inverness Journal* reported on New Year's Day 1808, that 'it was hinted that other articles had been found ... a ball and bar also of silver; but the labourers kept the matter a profound secret, as steps had been taken to compel the owner of the chain to deliver it up to the Crown.' Clearly, the Crown succeeded, as the chain can now be seen in the National Museum of Scotland.

At Loch Lochy, he was back marvelling at the mass of earth which 'had been thrown up on both sides along the whole line, that the men appeared in the proportion of emmets [ants] to an ant-hill, amid their own work' (Southey, 1929). Such observations almost make up for the regrettable absence of an artist like John Cooke Bourne, whose wonderful engravings document the construction of the first railways in England. Southey was able to supply particular details too. 'The hour of rest for men and horses is announced by blowing a horn; and so well have the horses learnt to measure time by their own exertions and sense of fatigue, that if the signal be delayed five minutes, they stop of their own accord, without it' (ibid.).

For Southey, the work was 'the greatest … of its kind that has ever been undertaken in ancient or modern times.' At Fort William he observed, 'Here we see the powers of nature brought to act upon a great scale, in subservience to the purposes of man … A panorama painted from this place would include the highest mountain in Great Britain, and its greatest work of art … the greatest piece of such masonry in the world, and the greatest work of its kind, beyond all comparison' (Southey, 1972).

Further south, Southey was less impressed by Wade's bridge at Aberfeldy – 'creditable neither to the skill nor taste of the architect' – but his judgement may have been clouded by the state of the village, which he called 'Aberfilthy'. He was thrilled to be in the first wheeled carriage ever to reach Strome Ferry, and was impressed by the quality of the Struie road

The bridge at Aberfeldy, built by General Wade.

General Wade's road leading out of Aberfeldy has all but disappeared.

Many travellers at that time made a point of visiting Neil Gow, the famed violinist. This portrait appears in Thomas Garnett's tour in 1800. 'He favoured us with several pieces of Scotch music. He excels most in the strathspeys…but he executes the laments…with a great deal of pathos.'

which leads up and over to Bonar Bridge – 'nor is there anywhere a finer specimen of road-making to be seen' (Southey, 1972).

The roads of the north were certainly superior to those that were found in Perthshire, where some travellers 'found themselves on a sudden upon a sort of Devil's bowling-green, and cried out in alarm "What's the matter?" – supposing that the horses had got out of their proper line. But the driver coolly answered, "Perthshire – we're in Perthshire."' The northern roads, which previously were in poor condition or simply did not exist, were suddenly the best in the country. Southey described how they were made. 'The plan upon which he [Telford] proceeds in road-making is this: first to level and drain; then, like the Romans, to lay a solid pavement of large stones, the round or broad end downwards, as close as they can be set; the points are then broken off, and a layer of stones broken to about the size of walnuts, laid over them, so that the whole are bound together; over all a little gravel if it be at hand, but this is not essential' (Southey, 1972).

The benefits were considerable. The postmaster at Dalmally told Southey that where before carts could only travel with nine cubic feet of timber, some carts could now manage 20 or 21 cubic feet. A farmer thought that his sheep would be worth a shilling a head more because of the ease of transporting them. Not everyone was happy, though. The blacksmith, for instance, was losing £7.00 a year on coach repairs. For some others, the new roads made no difference at all. Neil Gow, the famed fiddle player, told Southey 'They may praise your 'braid' [broad] roads that gains by 'em: for my part, when I'se gat a wee droppy at Perth, I'se just as lang again in getting hame by the new road as by the auld one' (Southey, 1972).

On the whole, most people were amazed to find such good roads in this isolated region. 'Often', reported the *Penny Magazine* in 1837, 'the most remote and wildest part[s] of the country are traversed by a road as smooth and level, and not so much frequented or worn, as that which runs through an English gentleman's park.' The eighth official report of the Commission, dated 1 March 1817, expresses great satisfaction with their work. They are delighted with the road that connects Applecross, Loch Carron, Loch Alsh, and Kintail, noting that the line 'is extremely rugged, and the rate of expense beyond example in our Highland Road-making.' The work on the Tongue road is nearing completion – 'we shall rejoice', they say, when it is finished. But their proudest achievement was in 'the great Roads from the Northern Coast of Sutherland, and from the Northern Coast of Caithness, forming a junction at Bonar Bridge, and thence proceeding without the intervention of a single Ferry, towards the Capital of the Kingdom.'

The Commission admitted one failure, however, regarding the link from Portinlick (near Bonar Bridge) to Ullapool via Assynt. 'There has been no disappointment in our transactions, which We so much regret, as not to have been enabled to make this Road, in our apprehension almost the only one of general importance not now opened or about to be opened in the Highlands of Scotland by means of the Road and Bridge Act' (8th Commissioners' Report, 1817). The route had been surveyed, but presumably they had run out of money.

As with the other reports, this one is accompanied by a fine map by Aaron Arrowsmith showing all the roads, both new and old. The new roads stand out with their red tinting: one follows the east coast, and another goes through the centre of Sutherland up to Tongue. It is evident that there is nothing new branching out to the north-west, and it should be noted that not all these roads marked on the map actually existed. MacCulloch was hugely frustrated when trying to follow the road from Tongue to Thurso. He complained that 'this was very much like the rest of Mr Arrowsmith's Highland roads, which intersect the country in all directions as if they had been surveyed by Telford and paved by MacAdam, but were never seen or heard of by mortal man except the apprentices who work at his long table in Soho Square' (MacCulloch, II, 1824).

It was to be another 30 or 40 years before a complete network of roads in the area had been constructed, the work for each one set in motion by the local landowners. Osgood Mackenzie describes how the road at Loch Maree was funded by £10,000 raised by his mother as well as some government grants, thereby providing much-needed employment during the potato famine of 1846 to 1848. Osgood, a young boy at the time, was given the privilege of digging the first spadeful. 'How well I remember it, surrounded by a huge crowd, many of them starving Skye men, for the famine was more sore in Skye and the islands than it was in our part of the mainland. I remember the tiny toy spade and the desperate exertions I had to make to cut my small bit of turf; then came the ringing cheers of the assembled multitude, and I felt myself a great hero' (Mackenzie, 1949).

A map showing the roads of Telford (dotted lines) and Wade (solid lines).
From The Lives of the Engineers *by Samuel Smiles, 1862.*

Ironically, given the hardship he caused by the clearances from the inland glens on his estate and the bad name he acquired as a result, one of the greatest benefactors in respect of road improvements was the first Duke of Sutherland, George Leveson-Gower (1758–1833). His motives may not have been entirely altruistic, for he was keen to maximise the efficiency of his land. In addition to 32 miles of road and three inns, all built 'wholly through the instrumentality of His Grace' (*New Statistical Account*, 1845), he commissioned a fine, detailed map of the county from his surveyors Gregory Burnett and William Scott in 1833. The New Statistical Account for Assynt praised all the improvements 'attributable, in a great measure, to the Noble proprietors, and, were there no other benefit conferred upon it, Assynt, on this account, owes a lasting debt of gratitude to the late excellent Duke of Sutherland.' Perhaps his greatest contribution was the road over the Moine, the boggy tract of land between Hope and Tongue that had caused so much trouble and fatigue over the years. Anderson's Guide welcomes this new highway which had been completed in 1830. 'The expense attending the construction of this piece of road must have been very great, from the mossy nature of the ground: the foundation was formed of bundles of coppice wood, laid in courses across one another, a layer of turf was next placed over these, and the whole being covered with gravel forms a road of the best description. Great ditches and numerous smaller drains are excavated on either side to contain the moss water' (Anderson, 1842).

Today, as you drive from Tongue towards Hope along this road (which has recently been considerably improved), a lone house appears on the skyline on the left (see colour section, page x). It has now lost its roof, and contains an alarming amount of multi-coloured graffiti on the inside walls which seem particularly incongruous in this remote area. This is Moine House. It displays a barely legible stone tablet on its eastern wall explaining that 'This House, erected for the refuge of the traveller serves to commemorate the construction of the road across the deep and dangerous morass of the Moin….' The Duke has never been forgiven for his decision to clear the glens, and the house, a monument to a work of great labour, has been allowed to fall into a state of disrepair.

The far north remained comparatively unexplored by tourists throughout the nineteenth century, but over the rest of Scotland, the infrastructure was at last in place to welcome the new leisure classes. Up they came, armed with their shotguns and fishing rods, their traveller's guides – Black's, Murray's, Anderson's – all seeking to experience the vision created by Sir Walter Scott, and encouraged by the presence of the most distinguished tourists of them all – the Royal Family.

CHAPTER SIX

THE TOURISTS RUSH IN

So for ever will I leave
Such a taint, and soon unweave
All the magic of the place –
'Tis now free to stupid face
To cutters and to fashion boats
To cravats and to Petticoats.

(Keats, after his visit to Staffa)

The improvements begun in the Highlands in the early nineteenth century continued apace. The development of canals and steam power gave the new leisure classes opportunities that Glasgow, in particular, was quick to exploit, with steamship companies offering tours along the west coast, and up through Telford's new canal highway to Inverness. 'The mob has rushed in', writes MacCulloch. 'Barouches and gigs, cocknies and fishermen and poets, Glasgow weavers and travelling haberdashers, now swarm in every resting place and meet us at every avenue' (MacCulloch, I, 1824). Railway construction began in the 1840s; by 1858, it was possible to travel all the way up to Inverness via Aberdeen, where the passengers arrived at a station designed by Joseph Mitchell. In 1865 a link from Perth to the north was completed, cutting through classic Highland country – passing through Pitlochry, Blair Atholl, and Aviemore. Beyond Inverness, the line reached Thurso in 1874.

John Dixon, whose book *Gairloch and Guide to Loch Maree* was published in 1886, watched tourists – a gentleman and three ladies – arriving at Achnasheen station on the Lochalsh line. The day was wet and murky. They had arranged to be met by a carriage and pair so that they could visit Loch Maree. Later, Dixon came upon them at Kinlochewe Hotel. 'They told me that although the day was too wet for them to drive down to the shore of the loch, and too misty to admit of its being fully seen from a distance, yet they were quite able to say that they had seen Loch Maree, for at one point they had put their heads out of

The "Lovat" locomotive, no. 77. This operated on the Highland Railway between 1886 – 1923. A photograph by E. Pouteau.

The "Dunrobin" Steam Locomotive, which pulled the Duke of Sutherland's private train, parked here appropriately underneath his large statue. Used on the Far North Line to Golspie which opened in 1871, it was sold to the Highland Railway Company in 1895.

"All we saw of the Caledonian Canal". A cartoon by W. Ralston, published in an 1874 edition of The Graphic.

the windows of their carriage during a brief cessation of the rain, and had distinctly seen the water of the loch!'

The new visitors demanded guidebooks, which offered advice on what the traveller would need. Anderson's *Guide to the Highlands*, the first detailed aid to traversing the far north of Scotland, recommends soap, buttons, umbrella, spyglass, and, more ominously, 'medicine, chiefly laxative and sedative' (Anderson, 1842). It suggests avoiding the sheets at inns as they would probably be damp, and warned that, if refuge were sought in a shepherd's hut, it might be necessary 'to get straw or ferns as your mattress and next to them heather.' Walford's Scientific Tourist counsels,

"Loch Maree" by Augusta Keate, from an album of drawings by the Keate family. The father, Robert, was Surgeon to King William IV. Another Keate, John (Robert's brother) was headmaster of Eton. Art and drawing were important accomplishments in such families. Augusta may not have been very old when she drew this famous view down into Loch Maree from Glen Docharty.

'If the tourist is fond of good living, a bottle of fish-sauce put into his portmanteau will be very useful', adding that 'there are times and situations when the having taken some tea will be found agreeable' (Walford, 1818).

As for clothing, the 1837 *Penny Magazine* recommends that 'dress should have a tolerable share of warmth and substance, for rain is more prevalent than heat … [and] shoes must be strong and well-nailed … capped with double leathers on the toes … A Bramah ink-bottle and a few steel pens will be convenient, inasmuch as Highland public-houses do not abound in the best stationery; and a pocket compass is always a pleasant companion, and may be of essential service.' Alas, it could suggest no map 'sufficiently accurate to serve as a guide', the best available being that in Anderson's *Guide*, and another published by the SDUK (Society for the Diffusion of Useful Knowledge).

Added to the problems of navigation was the fact that advice from locals was notoriously unreliable. MacCulloch offers a typical conversation. He asks how long Loch Teachus (Teacuis) is:

> 'It will be about twanty mile' was the answer.
> 'Twenty miles! Surely it cannot be so much?'
> 'May be it will be twalve.'
> 'It does not seem more than four.'
> 'Indeed, I'm thinking ye're right.'
> 'Really, my friend, you seem to know nothing about the matter.'
> 'Troth, I canna say I do.'
> And this is a very common end of all enquiries in the Highlands … and I … am inclined to attribute it, in them, to civility, and good nature chiefly; rather than to pride, or a wish to avoid the appearance of ignorance respecting their country (MacCulloch, II, 1824).

Other travellers seemed to share the same experiences. Thomas Garnett welcomed the milestones north of Aviemore as they meant he would not need local knowledge. James Hall goes so far as to say that 'Highlanders are greatly addicted to lying…In their reports, they are apt to answer what is for their own interest, and what [is] agreeable to the person to whom the report is made, rather than the real matter of fact' (Hall, 1807).

The problems facing the nineteenth-century tourist were much the same as those of today. First, of course, the weather. 'Rain … is the great enemy of all Highland tourists' states MacCulloch (I, 1824). James Hogg quaintly talks of 'the boisterousness of the weather' (1888). At Loch Hourn, MacCulloch asked a girl how long the snow lay on the ground. 'Weel, I wot [know]' was the answer, 'It never gangs [goes] til the rain comes' (II, 1824).

The inns of Scotland were notorious. Double-booking frequently occurred as Dr Johnson found at Bernera: on approaching his bed, 'a man black as a cyclops from the forge' (Johnson, 1984) leapt out of it. Service was often slow, or, as the *Guide to the Highlands* puts it, 'the tourist's patience is sometimes not a little taxed by the tardiness of the attendance.' Worse,

"View from my window for the last month."

"How I intend going to work if this weather continues, having found it the only way to have dry stockings."

Two cartoons from J. Smart's Life and Troubles of ane Artist in ye Highlands of Scotland.

Potage de lait. – 1ᵉ entrée, au lait. – 2ᵉ entrée, du lait – entremets au lait. – Notre voyageur se croit en nourrice......

"Potage de lait, l'entrée au lait, entremets au lait. Notre voyageur se croit en nourrice," [Soup of milk, main course of milk, supplementary course of milk. Our traveller imagines he has a wet nurse]. Original caricature after the French artist 'Cham'.

'there is a lamentable inattention to cleanliness, at least in the staircase and passage, and about the doors' (Anderson, 1842). Housework was not an area in which Highlanders excelled (see colour section, page iii), in fact, they made a virtue of the lack of it: 'Dirt's luck' was the proverb. 'On removing from one house to another it was accounted unlucky to get possession of a clean house … If one, who was removing from a house, was jealous of a successor, and wished to carry off the good fortune of the house, the out-going tenant swept it clean on leaving it' (Gregor, 1874).

As you travelled north, the choice of food became increasingly limited. 'Oh dear!', wrote Keats. 'I must soon be contented with an acre or two of oaten cake a hogshead of milk and a Cloaths basket of Eggs morning noon and night when I get among the Highlanders' (Letter to Fanny Keats, July 1818; Keats, 1901). MacCulloch's diet in the far north was even more limited. 'No one need live worse than I have lived in Sutherland, on boiled salmon and oat cakes, and on nothing but boiled salmon and oatcakes, for weeks, at breakfast, dinner, and supper' (II, 1824).

Then there was the problem of communication. It was all very well learning a bit of Gaelic, but, even if you did, as Garnett's companion found, 'though he could ask several necessary questions, yet he could scarcely comprehend their answers; and when they had heard him once speak Gaelic, [he] could scarcely ever prevail on any of the persons whom [he] met, to speak any English' (Garnett, 1800). By 1837, the *Penny Magazine* was assuring visitors that 'no apprehension need be entertained by the traveller of any serious inconvenience arising from his ignorance of the Gaelic language. At the poorest and most remote houses of entertainment he will be sure to meet with somebody able to understand his wants' (April 1837). Meanwhile, the inhabitants saw their language being corrupted. The Reverend James Russell observed that 'some young men, indeed, who have received a smattering of education, consider they are doing great service to the Gaelic, by interspersing their conversation with English words, and giving them a Gaelic termination and accent. These corrupters of both languages, with more pride than good taste, now and then, introduce words of bad English or of bad Scotch, which they have learned from the Newhaven or Buckie fishermen, whom they meet with on the coast of Caithness during the fishing season' (*New Statistical Account*, 1845). As for the visitors grappling with the Scots language, John Dixon begs them not to pronounce the word 'loch' like 'lock' – a mispronunciation that can still cause irritation today (1886).

Contrary to popular belief, travelling on the Highland roads had been quite safe for some time. Catherine Sinclair, the daughter of Sir John Sinclair who was the editor of the 1799 edition of *The Statistical Account*, encountered on her tour north in the 1830s 'neither beggars, pedlars, highwaymen, nor turnpikes … [and] therefore travellers might almost leave their purses at home without finding it out' (Sinclair, 1840). Seventy years earlier, Boswell had persuaded Johnson to leave his pistols behind. However, the road improvements brought dangers of their own – from vehicles. Joseph Mitchell spent many hours enduring coach journeys. 'Down we swept one hill, and the impetus brought us half up another. The quick turns were taken, sometimes within six inches of the stones placed to define the edge of the road, or the corner of a bridge; still, neither these, nor the bolting or kicking of some of the horses, nor the darkness of the night, diminished our steady pace of ten or twelve miles an hour' (Mitchell, 1971). For Sinclair, the worst came at the end. 'How unpleasant the final show off is, made by drivers and horses on entering a town! Their whole speed is reserved for the narrowest lanes, ill-causewayed, and perfectly paved with children, flocks of whom may be seen flying in all directions, at our approach, like a covey of partridges after a shot has been fired. We tilted over the streets and bridges of Inverness, at full career, grazing against carriages and people, while the horse became more and more excited, till at last with a rapid swing we turned a sharp corner, and wheeled up to the hotel, stopping with a crash as sudden as if a portcullis had fallen before us' (Sinclair, 1840).

Poor Donald Sage found himself on the very first coach to go north of Tain when returning from Aberdeen after the death of his step-mother. 'It was a rather dangerous mode of travelling, the danger arising, however, not from the state of the roads, but solely from the welcome given to the newly-started conveyance by all the proprietors, tacksmen, towns and villages on the line of the road by which the coach passed. Every one of them must treat the passengers, guard and drivers, with glasses of whisky, with which the drivers in particular so regaled themselves as at length to be totally unfit to manage the horses. The coach at different times therefore made many hairbreadth escapes from being overturned' (Sage, 1899).

The roads brought proper mail services which operated on a regular basis. There was a daily service from Golspie to Thurso. Another went from Golspie to Lairg twice a week, and the 'Diligence Gig' had room for two passengers. From Lairg, two more such vehicles would leave: one to Tongue, and the other to Assynt. Prior to the instigation of these deliveries, remote areas had to rely on the ever-dependable post-runners, who would do the journey in all weathers and seasons. Sir Hector Mackenzie employed 'Little Duncan, a bit of kilted india-rubber, who, with a sheepskin knapsack on his back to keep his despatches dry … left the west on Monday, got the 60 miles [to Dingwall] done on Wednesday, and returning on Thursday delivered up his mail … on the Saturday and was ready to trip off east next Monday' (Dixon, 1886). Osgood Mackenzie remembers the last post-runner, Iain Moram Posda ('Big John the Post') who 'trudged, I might say climbed, through the awful precipices of Creag Thairbh (the Bull's Rock) on the North side of Loch Maree, passing through Ardlair and Letterewe, and so on at one time to Dingwall, but latterly only to Achnasheen.' 'Big John'

"Bound for the Moors". The transport from Lairg to Assynt. From The Graphic, *1875.*

"Highland Foot-Post". From The Highlanders at Home *by James Logan with artwork by R.R. McIan, 1848 (from the 1900 edition).*

fully lived up to his name: 'On one occasion a Post Office overseer from London, who was … following Big John on foot, fainted en route, and Big John managed to carry the fat official on top of the mail bag for several miles till he reached Ardlair' (Mackenzie, 1949). When the regular mail services started, Big John emigrated to Australia.

By 1827, Mitchell reported that there were seven different stage coaches passing daily through Inverness in the summer months. The journey from Edinburgh to London in a four-horse coach was down from 12 days to a mere 45 hours, while the trip from London to Thurso (783 miles) was accomplished in precisely four days and 50 minutes! The coach drivers had plenty of time in which to bond with their passengers. The anonymous writer from Dover of a travel account held in the National Library of Scotland was thoroughly entertained by John Campbell, 'a complete model coachman … a man of extraordinary natural abilities, and full of fun and anecdote … [who] kept us in a roar of laughter by relating anecdotes told with the most inimitable humour' (NLS 9233). On the other hand, Joass, a driver Mitchell knew, 'could solve any problem in Euclid, to the astonishment of the Oxford and Cambridge tourists' (Mitchell, 1971).

With all the other changes that were occurring in the region at this time came a shift in the attitude to the scenery. Where before the traveller was relieved to leave behind 'the

A group of worthies, almost certainly at The Edinburgh Club, 1870. The gentleman in the kilt on the right is Sheriff Alexander Nicolson. He was the first to climb the highest peak of the Cuillin Hills on Skye – it still bears his (Gaelic) first name, Sgurr Alasdair.

gloomy precipices … and the savage rudeness of the mountains' (Hall, 1807), the modern tourist was actively seeking them out. Charles Lessingham Smith was congratulated by Duncan Macintyre on his first ascent over the Cuillin with the remark, 'I often think in my own mind that it is very strange you noblemen should come to see these wild hills of ours, and our noblemen should go to London to ruin themselves; but you've the best of it, Sir, for you gain health and strength, and our lords lose both that and fortunes too' (quoted in Cooper, 1979).

It was, however, not only scenery that the new tourists sought. Queen Victoria made her trip down the Crinan Canal almost exactly 100 years after Culloden. The Hanoverian government at that time had tried to destroy the spirit of the Highlands; now it was that very spirit that was alluring to the inquisitive traveller. A simpler, more traditional and less materialistic way of life contrasted markedly with the huge changes that were going on in the more industrial south. Some of those that came to Scotland could sense that this uncorrupted existence was beginning to vanish forever. MacCulloch on Ben Lawers met a young shepherd boy, and 'I asked him to accompany me, for the sake of conversation, and, when about to part, offered him a shilling. This he refused: but it was forced on him, and, in so doing, I am sure I did wrong; for it is likely that he will never refuse one again, and will possibly end by demanding five' (I, 1824). Some 30 years later, the anonymous tourist from Dover was enjoying a night in Oban at the end of his trip. 'A smart looking boy, barefooted as usual, accompanied by four or five smaller boys now came up and offered to sing an English song: receiving our permission he gave us "The Gypsie Laddie". The boy really sang very nicely and we thoroughly enjoyed

Oban was a favourite resort for the Victorian tourist, with excellent rail connections to the dockside, as can be seen in this photo by Valentine (c.1885).

the song of "The Jolly Shilling" which he gave with great spirit, the younger urchins joining in chorus until the hills rang again, attracting numerous parties around us; we afterwards said it was a scene. The lads got so liberally rewarded, that I venture to say that it might not be the last song they will sing in the twilight on the hills of Oban' (NLS 9233).

Most of these early nineteenth-century tourists headed for the more southern beauty spots – Loch Lomond, or Loch Katrine for example. But in 1803, the year in which Telford received his authorisation from the Government, two travellers set off on separate journeys with every intention of reaching the far north. On 27 May, James Hogg writes, '[I] dressed myself in black, put one shirt, and two neckcloths in my pocket; took a staff in my hand, and a shepherd's plaid about me, and left Ettrick on foot, with a view to traversing the West Highlands at least as far as the Isle of Skye' (1888). In fact, his pockets were also stuffed with letters of introduction to men of note as far away as Sutherland. His aim was to find a farm in the north – this was his third such trek in two years. It was still early in his illustrious writing career. He had done some work for the *Edinburgh Magazine*, and been recruited to collect ballads for Scott's *Minstrelsy*. In connection with this, Hogg had met the great man in 1802, and his *Tours* as published are in the form of letters to Scott.

A little earlier, on 15 April, James Hall had embarked on what he called *A Tour of Scotland by an Unusual Route* (the title of his account published in 1807). Hall was a less celebrated author than Hogg. He seems to have come from Walthamstow in London, and was possibly curate of Radford Semele. He published a *Tour of Ireland* in 1813, and also some poetry. *The Scots Magazine* was disappointed by the account of his Scottish travels: 'We do not remember to have seen volumes so handsome in their outward appearance, and at the same time so destitute of all internal recommendation. They are a mere collection of gossiping and scandalous anecdotes' (August 1807). In fact, Hall's anecdotes make for delightful reading, though one or two details set off alarm bells. At the General's Hut, a small inn near the Falls of Foyer, which was said to have been General Wade's base when he was supervising his military roads, Hall describes taking a meal. He was offered oatmeal cakes, cheese and butter, 'when I observed, by my landlady's hands, that she had the itch [scabies]! Upon my asking her who baked the cakes, she replied "the maid." I asked to see the maid, who almost instantly appeared, and I found by her hands ... that she was infected also. Thus, from the idea of the mistress having made the cheese and butter, and the maid the cakes, I was so disgusted that I could not eat any more' (Hall, 1807). This had been exactly the experience of Alexandre de la Rochefoucauld and his companion in 1786; they had been appalled by the filthy state of the place, and they noticed that the whole family suffered from scabies. However, one has to admire their fortitude: 'Good travellers feel challenged rather than discouraged when they are exploring enthralled ... One has to eat, but we couldn't face the oatcakes' (Scarfe, 2001). Of course, it is perfectly possible that the same family, with

"Falls of Foyers". An engraving by T. Allom.
A sight not to be missed on the
Victorian tourist itinerary.

the same problem, was still running the establishment 17 years later, or it may have been a common problem, but the thought occurs that the good Reverend Hall, sitting in his house in Chestnut Walk, Walthamstow, concocted his *'Tour'* from other people's experiences – a minor Marco Polo.

To add more doubt to his narrative, the detailed account of his travels up the east coast suddenly becomes very sketchy indeed along the north coast. When in Thurso he announced that he 'felt a strong desire to see Cape Wrath', and set off, finding the road 'the most dismal and dreary… that even my horse thought so; as he often wished to return. Indeed there are so many rivers and torrents to cross, so much bad road, or rather no road at all … and, with all, so bad accommodation … that I was led to think, were the British Legislature to enact, that delinquents from the parish of St Giles, in London, and other parts of the country, should be transported here, instead of Botany Bay, it would be an improvement in our code of laws.' And with that, he continues, 'When I arrived at Cape Wrath…'. There is no mention of Tongue, Ben Loyal, the Moine, Ben Hope, Loch Eriboll, or Durness – an unknown land to many of his readers. What could have been the most interesting part of his travels contains no details whatsoever of either the landscape or the hospitality he found in this remote region, and one begins to wonder whether he made the journey at all by land. Perhaps, finding the going very tough, he followed the advice that he might well have received from the locals, and took a boat along the north coast which dropped him off somewhere near Cape Wrath.

Be that as it may, there are many anecdotes in Hall's entertaining account that ring true enough. At Crieff he visited the spa of Pitkethley Wells (Pitkeathly Wells), after which, 'having drunk a pint of the waters, I strolled about before breakfast, reflecting on the indelicacy of both men and women, almost everywhere, in sight of one another, running constantly behind bushes and hedges: but it was not long before I completely sympathised with them…'. Then, proceeding further north-east, at Keith,

> I saw a number of people collected in the streets as if some accident had happened.
> Upon inquiry, I found it was occasioned by a woman having gone three times to

*Left: "Druid and Highlander". An early 19th century illustration by W. Davison.
The term Druid has a long history in Celtic regions for a professional man.*

*Right: "The Surgery" by Erskine Nicol. The artist spent most of his
life in Edinburgh, but taught in Dublin between 1845–1850.*

Doctor Dougall, to have a tooth drawn; and as often run out of his house, her tooth-ach[e] going away whenever she saw him come with his instrument to pull it out. Though an extremely good-hearted man, and always glad when he had it in his power to do good, he was so irritated, when he saw her running out a fourth time, that, halloing after her, and ordering her to be stopped, he followed her into the street; and, having, as it was dry, laid her down, there pulled out the tooth, and left her, with half a crown, to a person to take care of her.

Moving on to Grantoun (Grantown-on-Spey), Hall admired the dancing, noticing how mixed the company was. 'I found here honourable and right honourables, and people from thousands a year, to those that were not worth sixpence, all dancing and happy … [and] everyone seemed familiar and easy; yet, propriety of conduct was uniformly and strictly observed.' A few miles further on, he came across a landowner who delighted in the fact that a pair of eagles regularly nested on his land. 'As the Eagles kept what might be called an excellent larder, when any visitors surprized the gentleman, he was absolutely in the habit, as he told me himself, of sending his servants to see what their neighbours had to spare; and that they scarcely ever returned without something very good for the table. It is well enough known, that game of all kinds is not the worse, but the better for being kept a considerable time.'

At Nairn, Hall observed that there were 'two towns, and two different people, as the people that come from the country, and intend to speak Gaelic, live in one end of the town, and those that cannot, or do not intend to speak it, live in the other.' He continued on into Caithness, which 'has yet a most dreary appearance', and where he experienced 'one thing peculiar to Caithness: the gentlemen gently pinch the toes of the ladies with their own toes, by way of making love, under the table at dinner or supper. I was astonished at a constant treading on my toes, one night, which was repeated after many wry faces on my part. Next morning, having mentioned the circumstance to someone, I was let into the secret. My toes, by mistake, received a compliment not intended.'

Unlike James Hogg, who at this time was making his way up the west coast, the Reverend Hall was more interested in matters ecclesiastical than 'making love'. He remarks on 'the small, simple, and modest' congregation on the Island of May that do not face the preacher when he delivers his sermon, but rather 'turn about their backs, and in this position listen to what he says with the most reverend attention.' On the other hand, he was not impressed by the state of some of the churches, finding one on the banks of the River Aven [*sic*] which was used regularly but had no door. 'The truth is, calves, cattle, sheep, swine, &c. &c. lodge in it all night. In going to a kind of pew in it, I went over the feet, and positively in one place, up to the calves of my legs in cow dung.'

He was alarmed by the preacher at Inverkeith who, thumping repeatedly on a cushion, chose to preach on the text 'Thou shalt not seeth [boil] a kid in its mother's milk,' and vociferated against 'the abominable barbarity of those who were so void of feeling as to think of such a dish.' The Walthamstow vicar quietly points out that 'neither have the people of Inverkeithing any kids; nor if they had, would they ever think of seething them in their mother's milk.' Hall was also alarmed by the number of dogs taken to church. 'So many dogs being collected often fight, and make such a noise during public worship, as not only disturbs the congregation, but endangers the limbs of many.'

Perhaps the best description of the cacophony experienced at a Highland service is to be found in the *Memoirs* of Elizabeth Grant, who lived at Rothiemurchus between 1820 and 1827. She remembers that the precentor would call out a tune – 'St George's tune', or 'Auld Aberdeen', 'Hondred an' fifteen', and then proceed with the first line on the key note, the congregation repeating it. Line by line he continued in the same fashion, followed by the congregation, who managed in their enthusiasm 'serious severe screaming quite beyond the natural pitch of the voice, a wandering search after the air by many who never caught it, a flourish of difficult execution and plenty of the *tremolo* lately come into fashion. The dogs seized this occasion to bark (for they always came to the kirk with the family), and the babies to cry. When the minister could bear the din no longer he popped up again and touched the precentor's head, and instantly all sound ceased.' Everyone dressed in their Sunday best, the church in spring 'agreeably scented … with a decoction of the young buds of the birch trees' which the girls used as a hair lotion. The men all 'snuffed immensely during the delivery of the English sermon; they fed their noses with quills fastened by strings to the lids of their mulls, spooning up the snuff in quantities and without waste. The old women snuffed too, and groaned a great deal, to express their mental sufferings, their grief for all the backslidings supposed to be thundered at from the pulpit' (Grant, 1911).

The Church held a crucial place in Highland society. Parishioners would travel up to 15 miles on foot to attend services, the women carrying their shoes to keep them in good condition, slipping them on when they arrived. Catherine Sinclair noticed that 'in many more

Clockwise from top left:

"The Cottage Toilet". An engraving after Sir David Wilkie.

"Picturesque Ross-shire Highland Outdoor Sacrament". A Tuck postcard (c.1900).

"The True Doctrine. You'll all be D....d". A caricature thought to be by George Cruikshank, 1818.

Three communion tokens. Those wishing to take communion would need to have such a token. From left to right: Kinloss, Moray 1752; Durness Kirk Session, 1803; Kinlochbervie Free Church, undated.

remote places, the high road terminates at the church door' and visitors found that the horses they had borrowed would often slow down or stop automatically when they passed the kirk. At Beauly, she met 'swarms of pedestrians … hastening along the high road, to attend a Thursday sermon before the Sacrament in some distant parish, all so gaily dressed, that we conjectured they must be going to a wedding' (Sinclair, 1840). For the women, dressing up was a crucial part of the event. The *New Statistical Account* for Gairloch states that 'when a girl dresses in her best attire, her very best habiliments, in some instances, would be sufficient to purchase a better dwelling-house than that from which she has just issued' (1845).

Some of these Church events could involve vast numbers. Osgood Mackenzie remembers well the preparations for the outdoor communion, which took place every three years. Thousands would attend, though only a few actually took communion. Most came just to hear the preaching. One minister, after seating everyone, 'suddenly shouted, "I see Satan seated on some of your backs" whereupon several screamed and more than one fainted and had to be removed. None of your milk and water preachers!' says Osgood with relish. 'The sensational is alone of use' (Mackenzie, 1949).

Another who preached to vast numbers was James Alexander Haldane, who published a *Journal of a Tour through the Northern Counties of Scotland and the Orkneys* in 1798. Originally a naval officer, he became a lay preacher in 1794, based in Edinburgh, making extensive evangelical tours throughout the country. If his arithmetic can be believed, the numbers that turned out to hear him were quite remarkable: at Elgin, 700 in the morning, and 1,000 in the evening – 'the audiences were very attentive.' On 14 August in Orkney, there were 1,200 in the morning and about 2,300 in the evening. 'Many of the people appeared much affected and in tears.' On 18 August, he 'preached in the morning to upwards of 3,000, and in the evening to upwards of 4,000.' On 20 August, he claimed the numbers peaked at 6,000. Their efforts were blessed: they crossed the Pentland Firth four times without any difficulty. The captain of the boat, out of respect for the preaching party, ordered that if any of the crew should swear, he would 'receive corporal punishment, which … [was] occasionally carried into execution.' The weather too held out, with no rain, 'although sometimes the clouds had a lowering aspect.' However, things went less well when they returned to the mainland at the beginning of September. They were hoping to move on, but one of their party fell and injured his leg, which delayed them in Caithness for six weeks. They preached in Thurso 'to about 300 persons, who seemed rather unconcerned', and on 20 September at Mey 'to about 200 … [who were] very careless and inattentive.' The good weather had broken, and on 12 September they reported that 'the inclemency of the weather, from constant rain and the swelling of a river, prevented many people from coming to the meeting.' At last, on 12 October they re-entered Sutherland but found that 'few, if any of the people understood the English language' (Haldane, 1798). They were better received at Dingwall, but never ventured further west.

As the nineteenth century wore on, so the influence of the Free Churches increased in the Highlands, much to the regret of John Dixon. Children were not encouraged to go to the services, the people took no part, apart from primitive singing, and Christian Festivals were ignored. The Sabbath was very strictly observed, and daily worship expected in each household. 'An air of settled gloom on the faces of many of the people … intensified on the Sabbath Day' (Dixon, 1886). Sixty years earlier, Elizabeth Grant gave a rather more positive and delightful picture of faith in the northern regions. She thought 'there was no very deep religious feeling in the Highlands up to this time', but rather 'fairy legends, old clan tales, forebodings, prophesies, and other superstitions quite as much believed in as the Bible. The Shorter Catechism and the fairy stories were mixed up together to form the innermost faith of the Highlander, a much gayer and less metaphysical character than his Saxon-tainted countryman' (Grant, 1911).

The Reverend James Hall found hospitality in the house of an excise officer, somewhere in the vicinity of Cape Wrath (his account remains a little sketchy in these northern districts). He was surprised to find a newspaper which had been published only eight days previously, in London. News did not always travel that quickly in this area: 60 years earlier the Reverend Macdonald confided in his diary that it took one month for the news of Bonnie Prince Charlie's landing at Lochaber to reach Durness – barely 200 miles away. Soon James Hall was on his way back, again with no details of his journey, stating simply, 'I made the best of my way back to Thurso' (Hall, 1807). From there he visited Orkney, then crossed to the Western Isles, returning to the mainland at Fort William, thence back to Edinburgh.

At some point, his path might have crossed that of James Hogg, but neither mentions the other in his account. 'The Ettrick Shepherd', as Hogg is known, had been having an adventurous time as he proceeded up the west coast. At Loch Sloy, he remarks on the changes of fortune such a journey brings. One night he was 'in this hovel … in the midst of dying wives, crying children, pushing cows, and fighting dogs; and the very next day, at the same hour, in the same robes, same body, same spirit … [he was] in the splendid dining-room in the Castle of Inveraray, surrounded by dukes! lords! ladies! silver, silk, gold, pictures, powdered lacqueys, and the devil knows what.' He noticed the 'four proportionally large turrets' of Inverlochy Castle, but found the owner 'has four most elegant daughters whom

"Inveraray Castle" A view in Beattie's Scotland *by T. Allom. 1838.*

*"Loch Duich, from Ratagan". A Victorian photograph. The Five Sisters
of Kintail are at the head of this magnificent sea loch. Assuming Hogg's
companions were dropped off at Ardintoul, there was a long boat trip ahead.*

I confess I admired more than the four turrets of the Castle.' He tried to climb Ben Nevis,
'but the mist never left its top for two hours during my stay.' An unfortunate incident took
place at Shiel House, where he had to share a room, not only with 'a number of little insects
common enough in such places', but also with 'a whole band of Highlanders, both male and
female, who entered my room and fell to drinking whisky with great freedom' (Hogg, 1888).
Luckily, he had put his money and watch under his pillow, but in the morning he found
they had stolen his letters of introduction to the Sutherland landowners, thereby perhaps
depriving us of his thoughts and views of that county.

He did go further north, but only as far as Ross-shire. At Ratagan, he enjoyed much music
and dancing, hosted by Donald Macleod, and he then teamed up with the Reverend John
Macrae who was bound for Lochalsh. He took dinner with the Reverend, and was delighted
to find that 'the company … consisted of twelve, which, saving the old minister and I, were
all ladies; mostly young ones, and handsome.' His delight increased further when, on leaving
the table to continue his journey, he discovered that one of these fine ladies, Miss Flora
Macrae of Ardintoul, would be joining them with her aunt. When the party was ensconced
in the boat, 'there being a sharp breeze straight in our face … Mr Macrae spread his great
coat on the old lady and himself. This was exactly as I had wished it, and I immediately
wrapped Miss Flora in my shepherd's plaid.' Hogg does not record her reaction to being

A photograph by George Washington Wilson of Glen Shiel, which lies at the southeastern end of Loch Duich. Typical thatched cottages can be seen in the foreground.

wrapped in a garment that had been well worn over the previous few weeks, but as for him, 'though I was always averse to sailing, I could willingly have proceeded in this position at least for a week' (Hogg, 1888).

However, it was not to be. They landed at a point halfway up Loch Duich, only to find themselves thrust into the middle of a wedding celebration. Hogg seized the chance to join 'with avidity in their Highland reels', and made sure that he kept off the drink. By the time they had returned to the boat, the relationship with Miss Macrae seems to have been getting serious, but eventually the ladies were put ashore at their destination. Mr Macrae called out, 'Now, farewell, Miss Flora! Without pretending to the spirit of prophecy, I could tell you who you will dream of tonight.' Hogg confesses in his narrative, 'Considering of what inflammable materials my frame is composed, it was probably very fortunate that I was disappointed of ever seeing Miss Macrae again, as I might have felt the inconvenience of falling in love with an object in that remote country' (1888). Miss Macrae's ardour may well have been dampened by a severe attack of flu she suffered the next day as a result of this expedition.

Hogg continued further north via Loch Carron and Kinlochewe, ending up in a most remote spot, on the north side of Loch Maree. This was the home of Mr MacIntyre of Letterewe, who kept a well-cultivated garden, the 'turnips and potatoes, in drills as straight as a line.' Here Hogg might have had another amorous adventure, for a Miss Jane Downie was also staying there. However, being 'daughter to a respectable clergyman' and the recipient of 'a genteel education', one senses she kept a distance from the ardent shepherd. Nevertheless, she certainly made an impression as they explored together this most spectacular area. Whilst clambering over the rugged terrain, he just had time to admire 'a scene worthy of these regions … a lady of a most delicate form and elegantly dressed, in such a situation, climbing over the dizzy precipices in a retrograde direction, and after fixing one foot, hanging by both hands until she could find a small hold for the other … the wind which had grown extremely rough took such impression on her clothes, that I was really apprehensive that it would carry her off, and looked back several times with terror for fear that I should see her flying headlong toward the lake like a swan.' But let us leave James Hogg at Loch Maree, in a state of ecstasy at scenery 'being dreadful and grand beyond measure' (1888), for Miss Downie had urged him to visit the Isle of Lewis, and it was to Stornoway that he next headed. He was not, however, in the company of his plucky female companion.

Another significant visitor to the far north was William Daniell in 1815, as part of his circumnavigation of the entire coast of Great Britain. His account, published in eight volumes between 1814 and 1825, is notable for the large number of superb aquatints, depicting the various places he visited. Sometimes he employs a little artistic licence for effect, but they are always true to the spirit, and they form an important record. Much of his journey was done by sea – it was after all a *Voyage Around the Coast*. But he often went on shore, and wrote interestingly and entertainingly about the conditions and the life he found there. A boat, he said, is a 'floating dungeon, where the perpetual apprehension of danger is superadded to the irksomeness of confinement' (Daniell, 1978). He had an alarming dose of that danger when, making his way round the Ross-shire coast after leaving Tanera, he was caught by a torrential storm. Possibly unbeknown to Daniell, MacCulloch was in the vicinity at the same time, and he briefly described the scenario. 'It was indeed a fearful storm; for it caught us so suddenly with every sail set, that it had nearly laid us on our beam ends … Daniell passed under our stern in the commotion; and made a narrow escape of becoming somewhat more intimate with the sea than was necessary for his aquatintas [*sic*]' (MacCulloch, II, 1824). Daniell sought refuge in the Summer Isles, a group of islands northwest of Ullapool, whose name promised so much, but which never failed to disappoint. 'Alas!' he wrote. 'No spicy zephyr wafted over the waves … and, on nearer approach, the expectation of shady groves and verdant glades was utterly put to flight by the stern aspect of about thirty islets, of various elevation, rocky, and in general steril [*sic*], or presenting a precarious and stinted vegetation for the sustenance of a few sheep' (Daniell, 1978).

Eventually, Daniell managed to round More Head, and sail into Lochinver Bay, from where he was led on foot to the base of Suilven, one of the sandstone mountains that rise so distinctly from the undulating bed of gneiss that dominates the area. His guide was the factor of the Countess of Sutherland, a Mr Gunn, but such was the remoteness of the region they were visiting that Mr Gunn had to procure a local shepherd as a guide. Daniell noticed that 'the sheep that have ascended on its accessible side are frequently tempted, by the patches of herbage on its declivities, to situations from which they cannot release themselves; and there, when they become exhausted, they fall an easy prey to eagles that have eyed them while hovering around' (Daniell, 1978). MacCulloch had spotted similar behaviour at Cape Wrath, where shepherds told him that the sheep 'had thus been suspended between life and death for three years. It is part of the trade of these eagles to attempt to throw them down into the sea; an operation in which they often succeed, as I once witnessed in Skye' (MacCulloch, II, 1824).

Mr Gunn then conducted Daniell to an even more remote spot, Mr Charles Clarke's house which lay at the top end of Loch Glendhu, amidst magnificent wild scenery. 'The ride', wrote Daniell, 'to the head of Loch Assynt, though a distance of only four or five miles, was attended with some fatigue, in consequence of the rough and broken road, which presented every variety of ground that can excite a traveller's antipathy.' Daniell was now in an area that

"Unapool in Kyles-cu, Assynt". An engraving by William Daniell, 1820. Quinag towers over the scene. A bridge now crosses the sea loch at Kylescu, from where the loch splits into two, with one arm, Loch Glencoul, continuing in a south-easterly direction, and the other, Loch Glendhu, due east. Clarke's house lay at the further end of Loch Glendhu.

"Kyle Sku, Sutherland. Sept/57". Indistinctly monogrammed AWE(?). A very scarce Victorian view. Visitors to this district were still rare, and one wonders if this is just a copy of Daniell's aquatint. However, the depiction here of Quinag is a good deal more accurate.

was unlikely to have been visited before by any mere traveller passing through. They took a boat on the loch. 'The place is rightly called Glendhu, or the dark glen; for it is situated at the narrow termination of the loch, where the rocks on either side, though not very lofty, totally exclude the surface of its waters from the sun during three months in the year.' It was here that they found Mr Clarke's house, which also failed to glimpse any sun for the same period. 'Even the glorious light of summer seems to greet these haunts with a reluctant and quickly-fading smile; and during the long and tedious winter months they are doomed to receive little more than an alternation of twilight and darkness' (Daniell, 1978).

Daniell was apprehensive of what he might find in so gloomy a situation, and therefore was amazed to discover the house filled with culture and life. Mr Clarke turned out to be 'a man of enterprise and activity … and his fireside is enlivened by a family of sons and daughters, who are in no want of agreeable and useful resources to make the time pass pleasantly away: books, music, and conversation afford them a variety of amusement, and the social circle is at times enlarged by the accession of a few occasional visitants, who never fail to experience a friendly and cordial welcome.' The Clarke family have farmed in the area for 200 years, a story told in Reay Clarke's *Farming in Sutherland*. It is a typical story of farming in the Highlands.

In the case of Charles Clarke of Glendhu, he over-reached himself, 'as did a good number of other flockmasters at this time' (Clarke, 2014). The end of the Napoleonic Wars saw the price of wool and mutton come tumbling down, and he was declared bankrupt in 1824. He had to leave Glendhu. Four of the children emigrated to Tasmania, one calling his farm 'Glendhu'. It is still there. So too is Charles Clarke's house, lying not far from the Stack of Glencoul. Visit it, and you will be walking on ground hallowed not only by the Clarke family, but also by geologists from all over the world.

Daniell continued round the north coast, visiting Smoo Cave near Durness and the Kyle of Tongue, where he found people gathering cockles on the large expanse of sand that was revealed at low tide. These molluscs provided crucial nutrition for families from all over the county at times of famine, and Tongue was well known for being an area where they could be collected. Daniell carried on along the north coast, taking in Orkney before he began his long journey south.

Visitors to these northern parts, like Daniell, were still few in number, even 20 years later, according to the *Penny Magazine*. 'Into that large region which lies to the north of the Great Glen a very small proportion of travellers is tempted to venture … this part of the Country has been little puffed up by guide-books, and therefore those who follow their guide-books know nothing of it' (March, 1837). However, the scenery that dominates the far north was

Left: *"Entrance to the Cave of Smoo". An engraving by William Daniell, 1820.*

Right: *"Bay of Tongue". An engraving by William Daniell. The house to the left on the far shore is Tongue House, from where Lord Reay watched the* Hazard *founder in 1746.*

Many of those that did travel to the far north were keen sportsmen. In these two fine sketches, the man resting in the left-hand painting is in the Valley of [Ben] Hope, while the mountain in the right-hand painting looks like Ben Gholach, or possibly Cul Mor. Both are dated 1847.

Left: An original watercolour, looking down onto the River Almond at Perth. Unsigned, but attributed to William Gilpin.

Right: Arthur's Seat, Edinburgh. An engraving from William Gilpin's highland tour which he undertook in 1776. He found the shape of the old volcano 'odd, misshapen, and uncouth.'

increasingly that for which the tourist at that time was looking. Symptomatic of the change in attitude was a rather self-conscious aesthetic movement that prided itself on seeking out nature in its rough, untouched form – what was called the 'Picturesque'. The headmaster and Anglican priest William Gilpin was one of the driving forces behind this outlook. He famously suggested taking a mallet to Tintern Abbey, which he thought insufficiently ruined. One of his several tours took him to Scotland, an account of which he published in 1789. However, the rather precious sentiments of the movement left the members open to ridicule, which found expression in William Combe's satirical *Tour of Doctor Syntax in Search of the Picturesque* which was published in 1809. Combe followed this in 1821 with *The Tour of Doctor Prosody in Search of the Antique and Picturesque, through Scotland, the Hebrides, the Orkney and Shetland Isles* with wonderful illustrations by Charles Williams

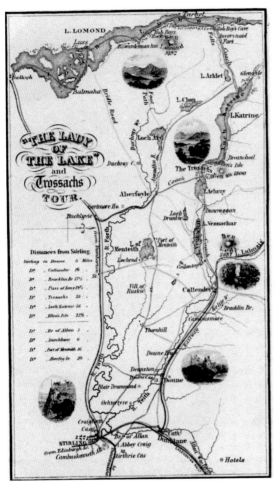

Map of The Lady of the Lake and Trossachs Tour. *Published by Robert Shearer, Stationer, Stirling.*

and W. Read (see colour section, pages xi and xii). Doctor Prosody may well have had fun in mocking Gilpin and his disciples, but there was no doubt that attitudes to the wild landscape of mountain and moor were, in general, changing.

There was another author at this time making an even greater impact, whose novels had quickly found a world audience. At first published anonymously, it was eventually revealed, apparently by Catherine Sinclair, that these were by Walter Scott. So powerful were his descriptions, and so in tune with the times was his writing, that people flocked to the areas that he depicted in his stories. His publisher, Robert Cadell, proudly declared that 'It is a well-ascertained fact, that from the date of publication of *The Lady of the Lake*, the post-horse duty in Scotland rose in an extraordinary degree, and indeed it continued to do so regularly for a number of years, the author's succeeding works keeping up the enthusiasm for our scenery which he had thus originally created.' Much of this scenery was further south – Loch Katrine, Loch Lomond, Midlothian – but the books brought tourists to these areas in their hundreds. Lord Cockburn in his diary of 1838 wrote that, 'The Inn near the Trossachs could, perhaps, put up about a dozen, or at the very most, two dozen of people; but last autumn I saw about one hundred apply for admittance, and after horrid altercations, entreaties and efforts, about fifty or sixty were compelled to huddle together all night. They were all of the upper rank ... but the pigs were as comfortably accommodated. I saw three or four English *gentlemen* spreading their own straw on the earthen floor of an outhouse' (Diary, 11 September 1838; *Circuit Journeys*, 1889). The anonymous journal-keeper from Dover was clearly a well-read Scott aficionado. At Perth he sought out the Church of St John ('scene of the ordeal in The Fair Maid of Perth'), then on to Killiecrankie ('the death of Bonnie Dundee'), and, eventually, Loch Katrine, where 'standing

*Left: Victorian sketch of Loch Katrine, signed indistinctly Julianna Dune(?).
Ellen's Isle, which features in Scott's 'Lady of the Lake' is in the middle distance.*

Right: Edie Ochiltree. A character from Scott's The Antiquary.
An original drawing by Mary Sealy ('MES'), 1844.

*Left: Two of the more remarkable travellers in
the Highlands at this time were John Bristed
and his companion. They chose to disguise
themselves as Americans 'because their
nation was a favourite with the Scottish'.
However, they simply drew attention to
themselves, and were arrested on suspicion
of being spies in Dundee. The account of
their travels is of limited interest, but it gets
a prize for its title:* Anthroplanomenos;
or A Pedestrian Tour through part of the
Highlands of Scotland in 1801.

on Ellen's Isle, one almost became a convert to the belief of the natives of the district, that the incident related in Scott's charming dramatic poem really occurred' (NLS 9233).

In this journal, the author also records that he met an American tourist who 'was travelling for the benefit of the son's health. The father with all the usual Yankee conceit, but any little unpleasantries in his bearing arising from this, in my opinion, completely redeemed by his solicitude for his son, which was pleasing to witness. It struck me as somewhat strange to see a *shopkeeper* from America, travelling in Europe for the sake of the health of an individual member of his family; only fancy an English *Grocer* travelling in America for a similar purpose. But the Yankees are a strange locomotive people, travel where you will: you are certain to find them' (NLS, 9233).

Sportsmen too were discovering the potential of the country. Osgood Mackenzie shot thousands of game birds up in Ross-shire, and then wondered why their numbers were decreasing! Earlier in the century, in 1824, a keen enthusiast who signed himself *Scotii Britannicus* went on a sporting tour to Caithness, choosing to travel much of the way by steamship, the party trusting their 'immortal bodies in those sailing engines of smoke, fire, and hot water … We all seem to forget the possibility of an *exit in fumo* [smoke].' They arrived at Glasgow hoping to board their vessel the Ben- Nevis. They were told 'Low Water, can't sail till 9 O'clock.' He continues 'We *look* at breakfast in a dirty Inn – the horn sounds

"Deerstalking. Halt for Lunch". A large group of sportsmen out in the Highlands, possibly Glen Tilt. A photograph by Valentine.

– we get on board – "Put On!" – 'tis the signal – the engine works, and we are deluged with a shower of filth from the flue.'

They arrived at the Crinan Canal at 10 p.m., having dined on board at 4 p.m. He found 'dinner the only *tolerable* thing about the vessel … [which is] now crammed full of people, and fifty brace of dogs: at all times top heavy, made much more so by an immense pile of baggage &c. upon the deck … I began to think the Ben-Nevis Steamer the worst built, commanded, and conducted vessel in the service, which certainly in reality … turned out to be the case.' At the Crinan they managed to get 'tolerable beds' at Lochgilphead, two miles away, the Inn 'not inferior to any English Country Inn', but full of people from another boat. Next day, 'after being wakened by the bagpipes, we walked eight miles to the other end of the canal, beating the steamer, which is always retarded by the locks and narrowness of the navigation.' They arrived at Fort William at 10 o'clock that evening, where they managed to secure 'good accommodation' thanks to a gentleman who knew exactly how to get a bed for the night. His sister failed to do so; she had expected to find a room at a new inn further on, but it was full and she had to remain on deck all night, 'all being disorder and insubordination on board, and [the] Captain intoxicated as usual.' Some men, unable to get beds, took 'a *trip* to the top of Ben Nevis … during the night'. He notes that there is still doubt as to whether Ben Nevis or Ben Wyvis is the highest mountain in Britain, but either way, most of this party returned 'all dead beat – [and] one was left on the mountain, quite knocked up' (Scotii, 1825).

They eventually arrived in the vicinity of Inverness at ten o'clock at night. It was the end of their voyage, and they were 'heartily tired of it, … resolving never again to enter the Ben-Nevis steamer … The confusion at the landing was dreadful – pitch dark – no lights to be got – all uproar – baggage tumbled about, and every one looking almost in vain for his own.' After an excellent supper at Mr Wilcox's Hotel, they continued their journey next day on the mail coach to Wick, 'a flourishing herring and salmon fishing village, though not so in cleanliness and order' (Scotii, 1825).

The most celebrated tourists were the royals and their entourage. George IV's trip to Edinburgh in 1822 was a key moment for Scottish cultural identity. In an effort to present the king with a strong picture of Scottish tradition, enquiries were made into tartans, and the clan system, which until then had meant little to the Lowland Scots. The research, which revealed that there was general ignorance about the subject, burgeoned into books like Logan's *Scottish Gael*, published in 1831, and McIan's beautifully illustrated *Clans of the Scottish Highlands* of 1843. With treatises like these, the Victorian concept of Scotland, all clan tartans, bagpipes, highland dancing, and highland games, was allowed to blossom into a culture we recognise to this very day.

George IV's visit also saved Elizabeth Grant's family, which was struggling financially. George IV would only drink the 'moonshine' [illegal] whisky, and great efforts were made to find a supply. 'My father sent word to me', wrote Grant, 'to employ my pet bin, where was whisky long in wood, long in uncorked bottles, mild as milk, and the true contraband *goût* [taste] in it … it made our fortunes afterwards … The whisky, and fifty brace of Ptarmigan

"Crinan Canal, 8 Aug. 1863". A very precise sketch by William Waterhouse, one of a number depicting his tour of Scotland.

"Ben Nevis". A detail from an original painting by William Donnelly, 1882. The Steamer is moored at Corpach, where a lock leads into the start of the western end of the Caledonian Canal. The town of Fort William can be seen in the distance.

Above: "Her Majesty in the Highlands – A Luncheon at Cairn Lochan, 1861". An engraving after Carl Haag, taken from The Graphic, *1880.*

Left: Title Page of The Clans of the Scottish Highlands *by James Logan, from sketches by R.R. McIan. It was published "Under the Patronage of the Highland Society of London." It was a key book in establishing the romance of Scottish history.*

Left: "A Royal Deer Stalking Party". An engraving after Carl Haag, taken from The Illustrated Times, *1858.*

Whether consciously or unconsciously, the image inspired a number of similar, if less dignified images on the same theme.

Right: "Our Trip to Ben Lomond". An original cartoon by William Donnelly, 1871. The artist, with his distinctive red beard, can be seen at work top right.

"The Queen at Loch Callater". From the Illustrated London News, *1880. Like many educated ladies of the time, the Queen was a very competent artist.*

all shot by one man in one day, went up to Holyrood House and were graciously received and made much of' (Grant, 1911). As a reward, her father was given an Indian Judgeship.

No one revelled more in this new-found 'Scottishness' than Queen Victoria and Prince Albert. They purchased Balmoral Castle in 1848, and magazines like *The Graphic* and *Illustrated London News* abound with engravings of the Queen picnicking, climbing, riding, and sketching in the Highlands, as well as presiding over dancing and sports at her Deeside home. Whilst on an expedition to Torridon, she came upon an unfortunate old man who was well known to the inhabitants of Gairloch. At one point he had taken his family to America, where most of them had died. The old man returned to Scotland and divided his time 'among those who are kind to him, – and they are not a few' (Dixon, 1886). Victoria noted in her diary, 'An old man, very tottery, passed where I was sketching, and I asked the Duchess of Roxburghe to speak to him; he seemed strange, said he had come from *America*, and was going to *England*, and thought *Torridon* very ugly' (Victoria, 1884). Such was her contact with her subjects in Scotland, always a little distant and aloof. But it was a country she loved for its 'poetry and romance, traditions and historical associations' (Cooper, 1979).

These were the very things that appealed to all the artists, poets, and musicians who felt drawn to the north. Wordsworth, as we saw earlier, came with his sister in 1803, though he was not always inspired to poetry of the highest order:

> Fort Augustus
> Did disgust us,
> And Fort William the same.
> At Letter Finlay
> We fared thinly;
> At Balachulish
> We looked foolish
> Wondering why we thither came.

Mendelssohn came 'for folksongs, an ear for the lovely, fragrant countryside, and a heart for the bare legs of the natives' (Letter, 1829, quoted in Jenkins and Visocchi, 1978). He produced a fine *Scottish Symphony*, full of the melancholy he clearly felt in the Highlands. He summed up his recollections thus:

> We wandered ten days without meeting a single traveller; what are marked on
> the map as towns, or at least villages, are a few sheds, huddled together, with one
> hole for the door, window and chimney, for the entrance and exit of men, beasts,
> light, and smoke, in which to all questions you get a dry 'No', in which brandy is
> the only beverage known, without church, without street, without gardens, the
> rooms pitch dark in broad daylight, children and fowls lying in the same straw,
> many huts without roofs, many unfurnished, with crumbling walls, many ruins
> of burnt houses; and even these inhabited spots are but sparingly scattered over
> the country … the rest is heath, with red or brown heather, withered fir stumps,
> and white stones … Fancy in all that the rich glowing sunshine, which paints the
> heath in a thousand divinely warm colours, and then the clouds chasing hither
> and thither! It is no wonder that the Highlands have been called melancholy.

And with memories like these in his head, it is perhaps not surprising that the supposedly positive coda to the symphony sounds a little forced and contrived. As for the *Hebrides Overture*, it is hard to know how much inspiration he gained from his trip to Staffa, as he was extremely seasick throughout the voyage. His German travelling companion, Karl Klingemann, comments that Mendelssohn 'is on better terms with the sea as a musician than as an individual or stomach' (Letter home, 10 August 1829, quoted in Jenkins and Visocchi, 1978).

Of all the nineteenth-century tourists who came to Scotland, no one can match the youthful enthusiasm and spontaneous delight of John Keats, who travelled up in 1818 with his friend Charles Brown. All his thoughts come from letters and poems, sent en route to his friends and family. 'I purpose', he told the painter, Benjamin Haydon, in a letter before leaving 'within a Month to put my knapsack at my back and make a pedestrian tour through the North of England, and part of Scotland … I will clamber through the clouds and exist. I will get such an accumulation of stupendous recollections that as I walk through the suburbs of London I may not see them' (letter, April 1818; Keats, 1901). Later in July 1818, he wrote to his friend Benjamin Bailey, 'I should not have consented to myself these four Months tramping in the highlands but that I thought it would give me more experience, rub off more Prejudice, use [me] to more hardship, identify finer scenes, load me with grander mountains, and strengthen more my reach in Poetry, than would stopping at home among Books even though I should reach Homer' (July, 1818; Keats, 1901). By the time the pair reached Inverness, he calculated that they had ridden about 400 miles, and walked more than 600.

Even before crossing into Scotland, Keats was delighted by the display of Scottish dancing he came across in Carlisle. 'They kick it and jumpit with mettle extraordinary, and whiskit and fleckit, and toe'd it, and go'd it, and twirld it, and wheel'd it, and stampt it, and sweated it, tattooing the floor like mad: The differenc[e] between our country dances and these scotch figures is about the same as leisurely stirring a cup o' Tea and beating up a batter pudding.' They visited Burns' cottage, where they met 'an old Man who knew Burns – damn him – and damn his Anecdotes – he was a great bore – it was impossible for a Southron to understand above 5 words in a hundred' (letter to Tom Keats, July 1818; Keats, 1901). They were also troubled by flies:

"The Shah in the Highlands: The Gillies' Ball, Glenmuick". From the Illustrated London News, *1889. The Shah of Persia was the guest of the Queen, and was introduced to facets of life in Scotland. The Ball followed a day of Highland Games at Ballater. The Shah looks a little bemused amongst all the dancing.*

All gentle folks who owe a grudge
To any living thing
Open your ears and stay your t[r]udge
Whilst I in dudgeon sing –

The gad fly he has stung me sore
O may he ne'er sting you!
But we have many a horrid bore
He may sting black and blue (letter to Tom Keats, July 1818; Keats, 1901).

They survived them, and the disappointment, too, of finding that the spot marked on their map 'Rest and be Thankful' was not the inn they were hoping for, but a seat and viewpoint, which meant they had to walk an extra five miles before getting breakfast. He barely survived the solo bagpipe, exclaiming, 'I thought the Beast would never have done' (letter to Tom Keats, July 1818; Keats, 1901). He could hardly believe that all the girls went around with bare feet, and was appalled by the hold the Church had over its congregations. 'These kirkmen have done Scotland harm – they have banished puns and laughing and kissing … I would sooner be a wild deer than a Girl under the dominion of the kirk, and I

"Rest and be Thankful". Glencroe. A traveller is resting thankfully in George Washington Wilson's photograph. The stone is a replacement for one put there by soldiers when building the military road in 1753. The climb out of Glen Croe is long and tiring.

"Inverness." An engraving by William Daniell, 1821.

would sooner be a wild hog than be the occasion of a Poor Creature's penance before those execrable elders' (letter to Tom Keats, July 1818; Keats, 1901). They were appalled at the cost of a trip out to Staffa, but managed eventually to find 'a boat at a bargain' (letter, 1818; Keats, 1901) which took them there. They sampled life in a shepherd's cottage – 'I am more com[f]ortable than I could have imagined' (letter to Bayley, 1818; Keats, 1901), though the floor was full of 'hills and dales', (letter to Tom, 1818; Keats, 1901). They also climbed to the top of Ben Nevis, Keats even ascending the cairn 'so got a little higher than old Ben himself' (letter to Tom, August 1818; Keats, 1901).

By the time the pair reached Inverness, Keats was in a poor way. He complained of a constant sore throat, and a doctor advised him to return south. 'The Physician here thinks him too thin and fevered to proceed on our journey', wrote Brown. 'It is a cruel disappointment.' Three years later, Keats was dead.

CHAPTER SEVEN

THE SCIENTISTS ARRIVE

Thou answerest not for thou art dead asleep. Is but
Thy life two dead eternities
The last in Air, the former in the deep –
First with the Whales last with the eagle skies –
Drown'd wast thou till an Earthquake made thee steep
Another cannot wake thy giant size!

(Keats, *Ailsa Rock*, 1901)

Whilst the tourist was lured north by the romantic image of a society untainted by progress and materialism, the scientist perceived a pristine wilderness latent with potential. In 1818, Thomas Walford published his *Scientific Tourist through England, Wales, and Scotland in which the Traveller is directed to the principal objects of Antiquity, Art, Science, and the Picturesque*. This was a time when the amateur could make serious contributions in areas such as botany, geology, and the antiquities, and much advice was offered in the introduction to this little book. There are six pages, for example, of detailed questions compiled by the Geological Society regarding veins, strata, angles, fissures, and fossils, followed by 13 pages on minerals, including such exotic examples as Black Wad, Scotch Corundum, and Toad-stone, together with more daunting ones like 'Lanmonite formerly called efflorescent zeolite' and 'Hydragillite or Wavellite'. There was also guidance for those searching for rare plants. Both mineralogist and botanist were advised to arm themselves with the following: 'a pocket compass and pocket barometer … and a small hand chisel and hammer to detach lichens and mineralogical specimens from the rocks. A botanical tin box for collecting plants and a quire of blotting-paper to arrange them in. A set of small boxes for curious minerals. A pocket telescope. A measuring tape or small walking-stick, an exact yard.' In addition, he adds, 'In hilly countries there is generally much rain, therefore an umbrella will be wanted.'

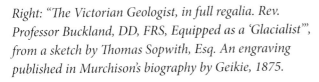

Above: A party of amateur botanists from Ross-shire, from a botanical scrapbook (c.1900).

Right: "The Victorian Geologist, in full regalia. Rev. Professor Buckland, DD, FRS, Equipped as a 'Glacialist'", from a sketch by Thomas Sopwith, Esq. An engraving published in Murchison's biography by Geikie, 1875.

Each county throughout Britain is described, with lists of antiquities to be visited, and minerals and rare plants to be found. Whereas Cornwall, for example, occupies 12 pages, Sutherland is covered in a page and a half. No minerals at all are listed, and as for flowers, 'In this mountainous county there certainly are a great variety of rare plants, although they are not mentioned in Lightfoot's *Flora Scotica* or any other botanical work in my possession. It will afford a rich harvest for some future botanist.' Lightfoot had been a member of Pennant's party in 1772, which had failed to penetrate deep into Sutherland.

In fact, the county had already been well served some 50 years earlier than Walford's book. James Robertson had made several forays into the far north-west between 1767 and 1771, even before Lightfoot. He had done much useful and ground-breaking work, having been sent there by the distinguished Dr John Hope, Superintendent of the Botanical Garden and Joint Professor of Botany and Materia Medica at the University of Edinburgh. Hope described Robertson as 'a young man, one of the gardeners, whom I had educated myself on a botanical expedition' (letter to Joseph Banks, 1767, quoted in Henderson and Dickson, 1994). At this time, botanists from Scotland were being sent all over the world: for example, Francis Meisson, Kew Garden's first plant hunter, and William Roxburgh, known as 'the Father of Indian Botany'. The north of Scotland was as unexplored as these more exotic locations, and Hope chose well in sending Robertson in that direction. His searches took him to the top of several mountains, crediting him with the first recorded ascents of, amongst others, Ben Wyvis, Ben Hope, Scaraben, Morven, Mayar, Lochnagar, Ben

Above: Ben Hope, a small painting by Arthur Perigal (c.1880).In spite of their magnificence, the mountains of the far north were sketched by few artists throughout the 19th century.

Left: Dr John Hope, an engraving by John Kay, one of his delightful caricatures of 18th century Edinburgh figures. Hope is seen here organising the establishment of the Edinburgh Botanical Gardens.

Lower left: "Primula Scotica", from David Wooster's Alpine Plants, *1874. Robertson originally listed this plant as 'Primula farinosa'.*

Avon, and Cairngorm. But his most celebrated 'first' was the ascent of Ben Nevis. Indeed, his fame today rests on that ascent rather than on his botanical achievements, though he made nothing of it, saying simply 'I ascended Ben Nevis, which is reckoned the highest mountain in Britain. A third part of the hill towards the top is entirely naked, resembling a heap of stones thrown together confusedly. The summit far overtops all the surrounding hills' (Journals, 19 August 1771; NLS MSS.2507-8).

His achievements in botany surely outweigh his ability to climb hills. On Bute in 1768, he listed 445 plant species – no one in Scotland had been anything like as meticulous as this before. Regarding Sutherland, a manuscript list survives of nearly 200 flora he recorded in the vicinity of Ben Hope and Ben More (probably Ben More Assynt). The new species he discovered included the curved sedge *Carex Maritima*, which he found at the mouth of the River Naver, and that most delicate and sought-after prize, *Primula Scotica*, which he called *Primula Farinosa*.

Robertson kept a detailed account of his travels, which fortunately has survived. At times they were clearly arduous. 'In examining this tract of desert, between Dydbol (Diebbiedale), and Strath Loikile (Strath Oykel) I wander'd for 3 days all alone till nightfall, scarcely knowing whither I went. The night which I was obliged to pass in one of the miserable huts, ill compensated the fatigues of my lonely strayings thro' the day. The Servant whom I brought from Inverness had returned home sick, and he who now accompanied me could not speak english, so that I was obliged, without an Interpreter to live among people whose language I did not understand' (Journals, June 1767; NLS MSS.2507-8).

Things improved when he found himself in the company of Mr Joseph Munro, Minister of Ederton, who led him to Loch Naver. Once there, Robertson climbed Ben Klibreck, and from its summit, he enjoyed watching a large flock of deer skipping along the brow of a hill below, but his view of these 'beautiful animals' was soon intercepted by a cloud of intervening fog (Journals, June 1767; NLS MSS.2507-8). Beautiful they might have been, but deer were a nuisance to farmers, destroying their corn. Various schemes were employed to control them. On Skye, Donald Nicholson told Robertson that 'a man shot three by successive fires without changing his station, while another man played on a bag pipe' (Journals, July 1768; NLS MSS.2507-8). The influence of music on deer was also noted by Mackinnon of Corryhatichan, who recounted how 'some nights when they came down to his corn, he played upon his Fiddle: during all the time he continued to play, they stood and heard him with the greatest attention' (Journals, July 1768; NLS MSS.2507-8).

From Loch Naver, Robertson journeyed up to Tongue, then on to Durness taking in Ben Hope on the way, thereby avoiding the Moine. He had hoped to explore many of the high hills in the area, but was plagued by bad weather. However, he was treated well by Lord Reay, who provided him with guides and provisions. In fact, wherever he went, he was met with extreme generosity; the ready hospitality of the inhabitants was 'a quality in which they are outdone by no people on earth' (Journals, June 1771; NLS MSS.2507-8). As a man of science, he had no time for their superstitions. He was shown adder-stones, which were pebbles with holes caused, he was assured by their owner Mrs Gordon of Fotherletter (Fodderletter), by snakes passing through them, thereby giving them healing properties. He later dismissed the idea in his diary as 'this absurd notion', but he had also to admire a toad-stone, a snail-stone and other similar pebbles, to all of which she 'ascribed great virtues' (Journals, June 1771; NLS MSS.2507-8). He was equally dismissive of the idea that 'some people tho' at a distance have the power of affecting milk in such a manner that it will yield neither butter nor cheese' (Journals, June 1771; NLS MSS.2507-8).

"Le Coup de L'Étrier chez les Montagnards Écossais". The generosity of Highlanders was well-known abroad. This engraving is from a French magazine of 1872.

He learnt of a well-known place of healing from his friend, the Reverend Munro, Minister of Tongue: Loch Monar (Loch Ma Naire) which was situated on the east side of the Naver, south of Skelpick. Donald Sage called it 'little better than a horse-pond' (Sage, 1899), but the 'cure' is mentioned in several sources, including the *New Statistical Account*, and the loch was clearly much frequented. The remedy was only effective on four days of the year: the first day of February, May, August, and November. Patients came from far and wide, even from Inverness and Orkney. They were required to be at the lochside by midnight, and were plunged into it three times soon after. They had to make sure they were away from the banks before sunrise. The Reverend Mackenzie says most were delighted with the treatment, but adds that they tended to be 'persons afflicted with nervous complaints and disordered imaginations, to whose health a journey of forty to sixty miles, a plunge into the loch, and the healthful air of our hills and glens may contribute all the improvement with which they are generally so much pleased' (*New Statistical Account*, 1845). Robertson was equally sceptical: 'They who know the power of imagination over the human frame, will not be surprised to hear that most of the plungers think themselves relieved' (Journals, August 1767; NLS MSS.2507-8). There were other sacred waters in the north, notably Loch Maree,

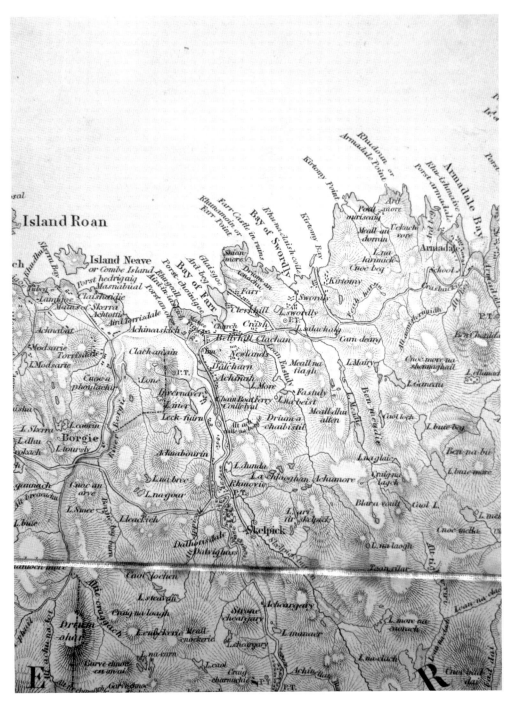

Loch Ma Naire, or Manaer as it is here on this detailed Sutherland map dated 1853.
A second edition of a map first issued in 1833 after a survey by Burnett and Scott
commissioned by the Duke of Sutherland. It was the first detailed map of the county. The
track from Skelpick leads straight to the small Loch Manaer (bottom centre).

where wretched 'deranged' people in the hope of a cure were dragged behind a boat as it toured the water round Isle Maree, in the middle of the loch.

Like other travellers, Robertson was unnerved by the apparent indolence of the inhabitants. 'Poor, ignorant, unskillful, unacquainted even with the appearance of fields properly cultivated, the inhabitants have sunk into a state of abject idleness, in which they continue not only from want of spirit, and of proper resources, but also from that veneration for old customs which impels the highlanders, like all Savages, to oppose all innovations however salutary' (Journals, June 1771; NLS MSS.2507-8). Wherever he looked, he saw the need for improvement. At Stonehaven, he advised farmers to clear their fields of 'supernumerary stones' (Journals, April 1767; NLS MSS.2507-8) and to use them to build walls, thereby clearing their land, and protecting it from the winds that were capable of removing the top soil overnight. On Skye he noted that 'they do not cut their barley, but pluck it up by the roots. Is not this custom very hurtful to the ground?' (Journals, July 1768; NLS MSS.2507-8). He was told they needed the length for thatching, but Robertson was not convinced. It troubled him to see so much ignorance and waste as he travelled down to Loch Broom, and then along the 'dreary road' to Contin, and back to Inverness.

On a more positive note, however, he was invited to join wedding festivities at Strath Dearn, near Inverness, which gave him great pleasure. 'The Bride was a buxom, blith widow of thirty, and seemed highly delighted with the prospect of quitting her weeds so soon.' Music was called for, but the fiddler had 'left his instrument on the other side of the water which was now unpassable.' No matter: another instrument was procured and the dancing 'cheered by repeated bumpers of Usquebough (whisky)' went on throughout the night until six in the morning, when they all sat down to breakfast. This consisted of 'a mess of the bowels of a sheep minced together, and swimming in butter. After this followed Curds and Cream, then Cheese, and last of all, Usquebough.' All this was taking place *before* the wedding ceremony itself. The bride and bridegroom were dressed in full tartan outfits (in spite of the Act of Proscription), and the father's permission was sought by the prospective husband at the door of the house with the words 'Are you willing to let me have your daughter?' It having been granted, they all set off for the church, the bridegroom with a sixpence in his shoe 'defending him from witchcraft' (Journals, June 1771; NLS MSS.2507-8). Guns were fired as they passed through villages, and over the bride's head as she emerged from the kirk.

There was a party later that night, in a barn. As she entered, the bride was showered with bread and cheese by her mother, but nothing was wasted, for 'these pieces were collected and eaten by the attendants.' Boiled mutton and broth were served for supper. 'The broth had no vegetables except oat-meal; and, a large piece of butter was put into each pot about ten minutes before the broth was taken off the fire, [which] covered the surface with oil.' And all the time the guests were served with 'a gumper [a large tot] of Usquebough.' Some of them had consumed at least two bottles of undiluted whisky, but at the end, the guests paid for everything they had drunk during the celebrations. The dancing commenced after supper, and continued until two in the morning 'with such graceful agility, as could hardly be surpassed even by the best taught Dancer.' The couple went to bed without ceremony.

Robertson's legacy following the various trips he made to the far north between 1767 and 1771 includes some fine botanical drawings which he undertook for Dr Hope, though many have been lost. After his Scottish travels, he went to sea as a surgeon's mate in 1772, and was promoted to Surgeon whilst in India, possibly through the influence of Sir Joseph Banks. His foreign travels gave him the opportunity to collect specimens from a variety of exotic locations, and he returned to England in 1789 with a modest fortune. He then married Helen Wilson in 1795, but his sojourn in India seems to have taken its toll on his health: he suffered from 'the gout and a bilious complaint', and 'took to opium with ardent spirits' (Edward Bruce, 1796, quoted in Henderson and Dickson, 1994). He died in 1796.

Robertson's memoirs display a knowledge of science that extends beyond botany. Science was not, then, such a specialised subject; for example, he was able to explain the spontaneous combustion that occurred at the coal pits near Brora which so 'surprized the ignorant spectators, but [would] not surprize an intelligent Chymist' (Journals, July 1771; NLS MSS.2507-8). It had been caused by pyrites which had decomposed in the air. He explained why Loch Ness never freezes – a feature that had caused much wonder and comment in the past. 'Any person who understands the process of freezing, will readily comprehend how the depth of a lake may prevent the formation of ice on its surface' (Journals, August 1771; NLS MSS.2507-8). As he travelled around, he noticed the composition of rocks, their strata, and direction of dip. His comments on such matters almost entitle him to be considered one of the earliest geologists to visit the area.

Geology as a science was still in its infancy in the early nineteenth century. William Smith's magnificent geological map *A Delineation of the Strata of England and Wales, with Part of Scotland*, published in 1815, is sometimes seen as the birth of modern geology. With it came, certainly, an understanding that there was wealth to be gained from an appreciation of the geological structure of an area. The need for coal, for example, was rapidly increasing as each year passed.

There had, of course, always been an interest in the composition and layout of the landscape, nowhere more so than in a country like Scotland with its array of different rocks and mountains. Much of the early science was mixed up with superstition and myth. MacCulloch, who was a central figure in early nineteenth-century geology, was irritated by observers like Martin Martin, whose *Descriptions of the Western Islands*, published in 1703, abounds with tales of giants dropping rocks that then form islands, and stones that when rubbed, cure diseases. 'It is to this ... gentleman that we are indebted for the tales of the two stones, which certain combating Giants threw at each other ... Mr. Martin, Mr Martin, you are nobody ...' (MacCulloch, IV, 1824). James Hogg, a great collector of Scottish folklore, in his novel *The Three Perils of Man* explains the creation of the Eildon Hills, near Melrose, as the work of a magician, whose spell divided one big mountain into the three hills we

Left: Dr James Hutton. A portrait by John Kay. The geologist is described as 'an ingeneous [sic] philosopher.' Hutton made important contributions to the science of geology, his investigations taking place further south in Scotland.

Above: View of the Eildon Hills, an engraving taken from General Roy's The Military Antiquities of the Romans in Britain, *1793.*

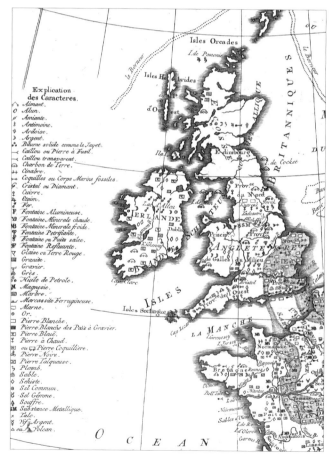

A detail from the Carte Minéralogique... d'une Portion de l'Europe. *By Philippe Buache, 1746. A very early geological map published for J.E. Guettard.*

see today. There may well have been times during the nineteenth century when scientists, struggling to work out the huge complexities of the landscape structures in Scotland, wished they could have fallen back on such simple solutions.

A very early geological map of part of Europe, published in France in 1746 by Phillipe Buache, shows a wealth of knowledge of France and southern England, which decreases as it goes further north. Scotland, as a whole, was thought to contain tin, lead, coal, iron, gold, and silver, but the only rock listed in the far north-west is marble.

The Highlands attracted a number of European geologists in the eighteenth and nineteenth centuries. One of the earliest was Barthélemy Faujas-de-Saint-Fond who visited England and Scotland in 1784. He is credited with being the first geologist to go to Staffa and recognise the volcanic origin of the distinctive basaltic columns found there. Trained as a lawyer in Grenoble, he rejected that profession, instead pursuing his interest in natural history in the Alps and Massif Central. His theories concerning volcanoes resulted in an important book *Recherches sur les volcans...* which was published in 1778. He would have heard about Staffa from, amongst others, Sir Joseph Banks, who had visited the island in 1772. On his arrival in London in 1784, Faujas was entertained at the home of the celebrated botanist, and later attended a rather alcoholic dinner and meeting of the Royal Society. His goal, though, was Staffa. As he set off from Glasgow, he realised that 'here, the traveller must bid farewell to English cleanliness: other manners and other customs now appear; but all can be borne when one is in quest of instruction' (Faujas, 1907). Indeed, there was much to be borne. After leaving Dumbarton, he and his party made their way to the inn at Luss, 'a single, sorry habitation by the side of the lake', where they discovered that a judge on circuit, bound for Inveraray, was occupying the only bed. It was pouring with rain, and the innkeeper, a formidable lady, insisted they tiptoe around so as not to wake the venerable lawyer. They begged to stay on account of the weather, but luring them back outside on pretence of assessing the rain, she slammed the door on them, and locked it, saying 'Be off!' Faujas was able to laugh at this, though it resulted in a 15-mile trek on to Tarbet. They arrived there at 3.30 am – 'never in my life did I make so disagreeable a journey', – but even then the troubles were not over, for it turned out that the one isolated inn was full of jurymen! However, this hostess was made of more generous stuff. She gave them not only a 'morsal to eat, and some excellent tea to warm [them]', but also two mattresses from her own bed, saying she had slept enough.

The problem continued to plague them, for the judge was expected at their next port of call, Inveraray, and they were informed 'very politely' that the Inn would not be able to provide accommodation. However, they carried letters of introduction to the Duke of Argyll, who received the party 'with every mark of friendship' and they ended up staying at his castle for three days. They even met the judge who had so bedevilled their progress; he came to dine one night, and was horrified to hear what had passed, assuring them that they 'should not sleep in the shed, if he had the pleasure of meeting [them] another time upon the road.'

The road from Inveraray to Dalmally was 'melancholy and painful … where during eight hours we met with no living creature, neither habitations, trees nor verdure.' (See colour

"Inveraray Castle, Scotland." A watercolour painting attributed to Henry W. Kerr. The castle is the seat of the Duke of Argyll.

"Intérieur de la Maison de Mac-Nab…" *An engraving by Le Brun that appeared in B. Faujas's* Voyage en Angleterre, en Écosse, et aux Îles Hébrides, *1797.*

section, page xii). However, Faujas was amazed at the 'elegance in so desert a place' of the inn at which they stayed. He was also taken with the sight of a group of men assembled nearby, in full Highland apparel, and who only spoke Gaelic. They had just come from the kirk, dressed in their finery. He noted that 'the English government, having repeatedly attempted to induce them to lay it aside, have never been able to succeed: though this attire is certainly least adapted to a people who live in so cold and humid a climate as this.'

Dalmally was also to show him the best and the worst of Highland hospitality. The innkeeper had found the ideal guide, Patrick Fraser, 'of modest and gentle manners … school-master of Dalmally' who could speak both English and Gaelic. Knowing Faujas's interest in matters of antiquity, Fraser was keen he should visit the cottage of the local blacksmith, named Macnab, who possessed items connected to the legendary figure of Ossian (whose epic poem published from 1760 is thought to be the invention of the Scottish poet James Macpherson).

Faujas was indeed keen to meet Macnab, and see inside his small cottage, though once there he was more than a little surprised, 'for it may be truly said, that the salon where the family waited for us, was in the chimney itself.' However, once he had become accustomed to the smoky atmosphere, he was able to pick out interesting features, such as the lamp consisting of 'lighted pieces of resinous wood, chiefly cut from the *Pinus taeda*.' It gave out an 'extraordinary light'. He was even more taken with the ritual, 'a sort of religious solemnity, arising from the desire of giving a kind reception to strangers.' A 'young, gentle and modest girl' whom Faujas presumed to be Macnab's daughter, shyly drank from a coggie [wooden bowl], and then offered it to her neighbour. 'It then passed from hand to hand, or rather from mouth to mouth, until every one had tasted it … We were then presented with butter, cakes made of oatmeal, and a little glass of whiskey. We returned our best thanks to this good family, who insisted on accompanying us back to the inn. Patrick Fraser informed us that it would be considered as an insult by these obliging people to offer them the most trifling gratuity.'

Unfortunately, Fraser was not at hand with advice at the next encounter a few minutes later. As they left the cottage, another Highlander approached, inviting them to pay his family a short visit, since they were ready to receive them. 'This man, who was richer and more ostentatious than Macnab, had made his wife put on her best finery … Her toilet, rather hastily put on, though not without some pretension, gave her an amusing air of embarrassment. She came up to inform us, that the fire was lighted, that the table was spread, and that the most excellent whiskey was poured out for us.' Not quite grasping the situation, Faujas and his party made their excuses, and thanking them profusely, headed for the carriages which were waiting. Too late, Fraser caught up with Faujas: 'You have painfully wounded the feelings of that family … by refusing to enter their habitation, while you visited that of Macnab. This kind of preference is regarded among the Highlanders as humiliating.' They quickly tried to retrace their steps, but 'the woman, on seeing us return, shut the door with a sort of temper … We were extremely sorry to find that we had given pain to people so hospitable and so polite.'

The next destination was Oban. When they were two hours away from the town, clouds began to form, and 'the rain soon fell in torrents; the darkness grew more profound, and in a few minutes it was no longer possible to see the road.' Patrick Fraser got out of the chaise, and attempted to lead them, feeling his way with his hands. They could hear the sea below them, and were advised to get out and walk for fear that the carriage would overturn and fall down a precipice. At one point it did turn over, but no damage was done, and, hopelessly lost, 'wet from head to foot, shivering with cold, and worn out with fatigue, we stood together around our carriages below some tall pines, shouting as loud as we could to any persons who might be within hearing to come to our assistance.' Not for the first time in such a situation, Faujas 'could not help giving way to a loud burst of laughter.' Their cries were not in vain: the local miller heard them, sought help from other locals, and led the party to Oban and to safety. They arrived at 1.30 a.m. 'Large fires were lighted to dry us, and after drinking a good deal of tea and some glasses of rum to warm us, we went to bed at four in the morning, and slept till ten.'

Rather than stay in Oban, his companions chose to head off to Mull, but Faujas wanted to explore the hills nearby. On the first evening he returned, delighted with the geological specimens he had gathered, but tired, and he went straight to bed. 'I had scarcely lain down when a cursed piper came and placed himself under my window. He waited upon me every evening in the passage of the inn, to regale me with an air; he then established himself in front of the house. There was no way of making him stop, and he went on to play this noisy instrument until eleven o'clock, with the wish to be agreeable to me, and to do me a kind of honour, of which I in vain endeavoured to convince him I was unworthy.' Faujas learned from Patrick Fraser that he was a distinguished piper, who wished simply to show his joy at his country being visited by a foreigner. 'I rose one evening with impatience, and not being able to make myself understood by speech, I took him by the hand to lead him to a distance. He returned immediately, however, as one who disputes a point of politeness, giving me to understand by his gestures, that he was not at all tired, and that he would play all night to please me; and he kept his word. Next day I forced him to accept again a small present, and made signs to him that I did not wish to hear him any more; but he was not to be outdone in civility. That very evening he returned, and made his pipe resound until midnight, playing constantly the same air.'

Other visitors had trouble with Pipers. This is from a series of cartoons depicting the travels of "Mr Aldgate-Lothbury in the Highlands." The Graphic, 1882.

"L'Isle de Staffa". An engraving by Durand from Faujas's account,
Voyage en Angleterre, en Écosse et aux Îles Hébrides, *1797.*

The only compensation for Faujas at Oban was the geology – 'several species and varieties of very curious rocks … [a] vast collection of different stones … deserves all the attention of those who love studies connected with the theory of the earth.' Mercifully, too, Staffa fully met his expectation, for he found it a 'superb monument of a grand subterranean conflagration … [that] presents an appearance of order and regularity so astonishing that it is difficult for the coldest observer … not to be singularly surprised by this kind of natural palace which seems a veritable marvel.'

Faujas returned to Edinburgh via Perth and St Andrews. As he entered the more gentle Lowland districts, he felt a sense of relief, such as one feels 'on the return of spring, though at this time we were at the end of autumn. But it may be said that all is winter, all is wild, dreary and sterile in the regions which I had just traversed.' Clearly, he had struggled in the Highlands.

Geologists from all over Europe were keen to visit Scotland. Some studied at Edinburgh University: the Swiss national Louis Albert Necker de Saussure, for example, between 1806 and 1808. He took the opportunity to travel, feeling drawn, as had Faujas, to the Hebrides, and like his fellow European, he experienced some trouble with the bagpipes. 'We were regaled with … music at Ulva House every day during dinner, and although the piper was placed outside of the house, it was almost impossible to hear the conversation.' Even inside Fingal's Cave on Staffa, he could not escape, though here he was more appreciative when the

"La Grotte de Fingal dans L'Isle de Staffa. Vue de l'Intérieur". An engraving by
L.A. Necker, *from his* Voyage en Écosse, et aux Îles Hébrides, *1821.*

"Women at the Quern and the Luaghad with a view of Talyskir". An engraving by Moses Grifith, from Pennant's A Tour in Scotland, *1772. There is a long tradition of working songs in the Highlands. MacCulloch came across women waulking (working the tweed or wool to shrink and soften it, as seen in this print) and describes 'coming suddenly on the bare-legged nymphs in the very orgasm and fury of inspiration, kicking and singing and hallooing as if they had been possessed by twelve devils.'*

piper 'made it resound with the wild and powerful notes of his bagpipe; this instrument well accorded with the character of the scene, and the notes prolonged by the echoes, produced an effect altogether analogous to that of an organ in pealing through the vaulted aisles of a vast cathedral' (Necker, 1822).

There was more musical entertainment on Iona, where they threw a party for the boatmen. The oar songs they sang were called *jorrams*, while those sung whilst working on the land were *Oran luathaidh*:

> The men and women seated themselves in a circle and joined hands, or held, in couples, the end of a handkerchief, with which they kept time during the chorus. Two of our boatmen, who were the leaders, made all kinds of grimaces and apish tricks whilst singing, striking themselves on the head one against the other with

all the dexterity of Italian buffoons, while the rest of the company were convulsed with laughter. This scene greatly amused us, and we were astonished to see, under so foggy an atmosphere, in so dreary a climate, a people animated by that gaiety and cheerfulness which we are apt to attribute exclusively to those nations who inhabit the delightful countries of the south of Europe (ibid.).

Necker travelled further north than Faujas, crossing from Skye to Loch Carron where he tried to find a guide to take him to Inverness. The first to offer his services wanted an exorbitant fee, and warned that he would have to carry the Swiss geologist on his shoulders through the fast-flowing rivers. Luckily, the carpenter of the village was able to offer a pony, with a young man as a guide, all at a reasonable cost. So they set off. Necker rode a horse no bigger than a goat, with a saddle made of old carpet. They passed through the familiar wild scenes of rivers, mountains and heath, safely reaching the east coast from where Necker then headed south to Edinburgh.

His travels were eventually published in three volumes in France, but not until 1821. Parts were later translated into English. He clearly felt at home in Scotland, calling Highlanders 'brother mountaineers', and in 1841 he settled in Portree on Skye where he led a rather solitary existence. He died there in 1861, and is buried in the churchyard.

Ami Boué is another European who, in his own words, 'went to Scotland to escape from Napoleon' (Geikie, 1904), such was the unsettled state on the Continent in the early years of the nineteenth century. Boué graduated from Edinburgh University in 1817 with a degree in medicine, saying of his time there, 'I wore out more umbrellas in my four years of residence in Great Britain than during all the rest of my life' (Geikie, 1904). He had certainly spent many hours out in the open, for geology was his real interest. In 1820 he published the first detailed book on Scottish geology, *Essai Géologique sur l'Écosse*, which Archibald Geikie described as 'a most valuable treatise which in many respects was far in advance of its time' (1904). The reviewer in the 1823 Edinburgh Review was less impressed, calling it 'a compound of plagiarism and conjecture.' The book is accompanied by what is considered to be the first geological map of Scotland (see colour section, page xiii), though it is admittedly a little low on details, especially in the Highlands. He had travelled around the country, but not, it seems, to the far north, for his remarks concerning that region are heavily reliant on a hugely important book by the geologist John MacCulloch published in 1819: *A Description of the Western Isles of Scotland including the Isle of Man*.

MacCulloch's name has already occurred frequently in these pages – it sounds Scottish, but he was in fact born in Guernsey. His mother was a native of the island, but he could trace his ancestors back to the MacCullochs of Galloway. He studied medicine at Edinburgh, and began his working life as a surgeon in the army, later becoming the Chemist of the Board

"John MacCulloch, M.D". An engraving after R.B. Faulkner, published in Geikie's Life of Murchison, *1875.*

of Ordnance. By 1814, he had been appointed geologist of the Trigonometrical Survey, and he was President of the Geological Society between 1816 and 1818.

In addition to the volumes on Scottish geology, MacCulloch published an account of his travels throughout the country in 1824, *The Highlands and Western Isles of Scotland*. If his *Description of the Western Isles of Scotland* is a classic of early geology, the later account of his journeys is an equally important addition to the travel writing of the time. It is a wonderfully lively and controversial picture of the Highlands in the early nineteenth century. It is a view formed not on one specific journey, but over a number of years.

'You will wonder, perhaps', asks MacCulloch, 'what I was doing in this country, so long and so often; why I explored the unexplored, why I risked my neck every day on mountain and precipice, and my whole carcase on flood and in ford; why I walked and rode and ferried and sailed, and hungered and watched, day after day, and summer after summer... It was Geology, my dear friend; Geology, divine maid. Did I not look with eyes of anticipation to the day when all my toils were to be rewarded by the display of the map of Scotland, with all its bright array of blues and greens and carnations and browns …' (MacCulloch, I, 1824). He did complete his fine geological map of the country in 1834, but it was not published until 1836. He never saw it in print, for, having survived all the dangers in Scotland, he died in a tragic road accident whilst on his honeymoon in Cornwall in 1835.

MacCulloch relishes a good story, and makes the most of every situation in which he finds himself. In so doing, he, like Johnson, made some enemies, including Mackinnon of Coirechatachan, whose family had previously entertained both Pennant and Dr Johnson. Following the publication of MacCulloch's travelogue, Mackinnon took the engraved portrait of the geologist to a crockery dealer in Glasgow and commissioned a set of 'earthenware' with MacCulloch's likeness on each. The 'earthenware' consisted of chamber pots, and the portrait was on the inside (Geikie, 1904)!

As a geologist, MacCulloch inevitably climbed various mountains, including Ben Nevis, with a guide whom he secured for five shillings, though in general his opinion of such people was low. 'The event did not belie my theory; for when my guide found himself in a whirlwind of fog and snow, so thick that we could scarcely see each other … he began to cry; lamenting that he should never see his mother again, and reproaching himself for having undertaken the office … He would even surrender his five shillings, if I would show him the way down the hill' (MacCulloch, I, 1824).

A photograph by the American visitor to Tongue in 1900, Emily Tuckerman. All her photographs are well annotated. On this one she has written 'Old Dolly in her little stone house – wee bit shawly over her head. She sat two minutes for this photo. Our dresses smelt of peat for months....We put a sixpence under a tea cup on leaving, and she thought the fairies had been round.'

He had better luck with his guide on An Teallach, a mountain he calls Kea Cloch, and which he believed to have been 'completely overlooked by map-makers and travellers.' His companion there was 'Honest John Macdonald ... a Highlander, with legs like a deer and muscles like fiddle strings.' At one point, he observed Macdonald 'perched, very much like [a] he-goat ... on the point of a projecting shelf.' They had come to a fast-flowing river tumbling down the mountainside. 'John talked of leaping the fall ... [and] in the turning of a handspike, he was on the other side ... I sighed and looked and looked again; but honour won the day, and I never struck the ground with half so much satisfaction.' This same John Macdonald later spotted a fairy, exclaiming 'Hey, what a bonny lassie ... [behind] that bush.'

MacCulloch continues, 'we ... beat all the bushes round ... but to no purpose ... After ten summers spent wandering among Highland hills and glens, amidst their mists and storms, in the very heart and centre of old romance, I have come away without knowing whether to believe in fairies, and other of the fraternity of Elves, or not: not doubting about my own belief, I should however say, but uncertain whether others believe.' He adds sadly that, 'we have ... been philosophised out of half our pleasures' (MacCulloch, II, 1824).

Some scenery he admired from afar. He might well have been sitting at Knockan Crag, which was later a key area in the Highlands Controversy and is now a visitor attraction in the North West Highlands Geopark, when he described most delightfully the distinctive hills that rise above the gneiss plateau in the Assynt region. These he felt were mountains 'which seem as if they had tumbled down from the clouds; having nothing to do with the country or each other, either in shape, materials, position, or character, and which look very much as if they were wondering how they got there' (MacCulloch, II, 1824).

He studiously avoided all geology in the accounts of his travels, though he was amused by 'the quarryman who accounted to me for a fissure in a rock, by the earthquake which happened at the Crucifixion, [which] shewed a bounding and philosophical spirit worthy of a seat among the geologists of the age.' On the other hand, he despaired of the posing amateurs – 'every blockhead who has cracked a stone at Salisbury craig, must display a hammer ... The world will never be the wiser for all their hammers' (MacCulloch, I, 1824).

Clockwise from top left:

*The Hull and Leeds Geological Society outing, 1889. The gentleman
on the far left is ready with his hammer.*

*"Ben Venue, Loch Katrine". A pencil drawing by J. Ewbank, dated 1833. After the publication
of Scott's* Lady of the Lake, *Loch Katrine became a major tourist destination.*

*"Parallel Roads, Glen Roy". A photograph by G.W. Wilson. Various
suggestions were offered to explain the 'roads', including that which saw them
as an ancient race track, before it was accepted that they mark the various
levels, or shorelines of a lake as it drained away at the end of the Ice Age.*

He delights in the unexplored regions he stumbled on, where 'human foot has
scarcely trod'. He believed himself to be 'the very first absolute stranger, to visit Loch
Cateran [Katrine],' which is not even marked on his map. While in the vicinity, he came
unexpectedly upon a 'young cockney apothecary ... who was employing the Edinburgh
summer vacation in practising on a French Horn. After three months of weary labour, he
had attained the fourth bar of "God Save the King", and the whole, valley, rock, mountain,
and water, resounded all day long with the odious notes and their more odious echoes.'

At Glen Roy, he not only admired the Parallel Roads, but correctly explained their formation (without fully understanding the process), doubting whether 'they have been seen by twenty people beyond those of their immediate neighbourhood.' He was refused accommodation in the vicinity 'because it was known that I was an unbeliever. Highland wrath must be powerful, so to overcome Highland kindness' (MacCulloch, I, 1824). He thought Loch Maree had never been visited by strangers, except for Pennant (he clearly had not heard of James Hogg's adventures), and at Coruisk on Skye he remarks, 'I suspect that the knowledge of this place is still limited to half a dozen persons ...' (MacCulloch, III, 1824).

On Eigg, he entered what he took to be the inn, where he confidently ordered a fire in his room, a change of clothes, and some dinner, before realising that the 'Innkeeper' was in fact a Mr Hardcastle, and that he was in the gentleman's house! 'I could only beg a thousand pardons, and conclude that he must think me a very impudent fellow.' Hardcastle laughed it off with typical Highland generosity. 'No! No! ... Bide ye quiet: we shall all be glad of your company' (MacCulloch, IV, 1824).

At Loch Broom, he passed another 'useless fishing establishment' (MacCulloch, II, 1824) – John Knox's dreams of thriving fishing communities had been dealt a fatal blow by the herring going elsewhere. On the other hand, MacCulloch thought the clearances were, in the end, a positive enterprise, taking the native Highlander from the glens, where he was 'a starving melancholy wretch, half-clothed, living in a dunghill ...' and transforming his existence on the sea shore 'where Lord Stafford [the Duke of Sutherland] has driven him, to the great annoyance of all romantic gentlemen.' Here, he has 'a comfortable cottage, with a boat, a cow, and a few acres of oats ... [and he is] active, industrious, and happy' (MacCulloch, I, 1824). Generally, he is critical of Highland indolence – indeed he devotes a whole chapter to it. He is amazed to find '... a piece of new-made road ... that ... passed through *the middle* of one of the ancient Highland huts ... the owner [having] received sufficient and regular notice to quit ... with the assignment of a new spot to build on. This, it appeared, was too much trouble; and, rather than be fashed with moving, he had remained quietly in his place, suffering the road-makers to beat his walls down about his ears, and then very peaceably repairing the damage ... There, I doubt not, he remains, and will remain, till the remnants of this structure fall upon his head. If you disbelieve me, go to Loch Inver, and see' (MacCulloch, IV, 1824).

Further south, MacCulloch's vessel pulled in at Poolewe, where he visited the local inn. He was clearly not impressed: 'amid a splendid crop of groundsel, chickweed, and docks, intermixed with grass and an occasional thistle, were seen two potato plants, and seven stumps, bearing just leaves enough to prove that they had once been cabbages.' As for the owner, Mr Mackenzie, 'three fourths of his time were spent in lounging about on the green before his door, for want of employment; and his elder children ... followed the paternal example.' MacCulloch cannot believe that Mackenzie fails, 'not only to eat his cabbages, but to make money off his rood of land, by selling to us and to other ships, what we would gladly have purchased with gold' (MacCulloch, II, 1824).

Flowerdale Bay, Gairloch. Ross-shire. A photograph by G.W. Wilson. MacCulloch would have been cheered to find this well-tended garden.

Highland goats. A photograph by Charles Reid (c.1890). Goats were frequently reared by the population of the Highlands during the 18th century, but became less favoured during the 19th century.

Certainly, the diet on board a vessel could be very limited. At Loch Broom, MacCulloch was disturbed in his cabin by a goat 'which the men had brought on board that we might be sure of milk for our breakfasts. Unluckily, when it came to be milked, it was discovered to be a he-goat; such was the pastoral knowledge of our boatswain' (MacCulloch, II, 1824). But he recommends doing the journey up the entire west coast, from 'Cantyre to Cape Rath' by boat, preferring 'the mountain wave to the mountain shore; a home on the deep, to the want of one among rocks and bogs, amid fords and ferries, through dub [pools of stagnant water] and learie [peat banks] and labour and starvation' (ibid.). Off Lewis, they enjoyed watching the bottle-nosed whales 'with their gambols and their spouts; but, on one occasion … the abominable uxorious beast had mistaken our boat for his wife, and made love so rudely that he had nearly knocked one of the men overboard with his tail' (MacCulloch, III, 1824).

However, his travels were not all non-stop action. As he lies becalmed in the Hebrides, he reflects,

Steam and Sail. "Loch Ness from Fort Augustus". A photograph by Valentine, 1891.

the sea stands still; the winds stand still; time seems to stand still … the mind takes the complexion of the elements, and falls into a gentle and dreaming kind of acquiescence; thinking nothing of the future, and scarcely knowing whether it is even thinking of the present … [Then] a long trailing line of black smoke in the horizon announced a steam-boat. She was soon up with us; and as she shot along under our stern, as if in contempt, our sails shook once more … the long column of smoke, diminishing to a point as she left us, spread over the clear sky, long remaining to mark the line of her fiery transit. The contrast was too great to be endured: it disturbed our patience; and two boats being put a-head, with the aid of a rising air and the tide, we reached Loch Ewe (MacCulloch, II, 1824).

MacCulloch, like Johnson, thought that life in the far north had already changed irrevocably, observing that 'the term Highlands is now, scarcely even a geographical distinction: the shade by which it unites with the Lowlands, is evanescent and undefinable; and, every year, the colours blend more, and the neutral tint widens around the border that once separated them. The term Highlander is still less definite … the Highlands have long ceased to form a nation and a people' (MacCulloch, I, 1824).

CHAPTER EIGHT

THE HIGHLANDS CONTROVERSY

*But if any human being desire to push on to new discoveries instead of
just retaining and using the old; to win victories over Nature as a worker
rather than over hostile critics as a disputant; to attain, in fact, to clear
and demonstrative knowledge instead of attractive and probable theory;
we invite him as a true son of Science to join our ranks, if he will, that,
without lingering in the fore courts of Nature's temple, trodden already by
the crowd, we may open at last for all the approach to Her inner shrine.*

(Francis Bacon, quoted in *The Centenary of the Geological Society of London* 1907)

During the nineteenth century, the science of geology made enormous advances. The
Industrial Revolution demanded the supply of a variety of minerals, the most obvious of
which was coal, and with this demand came a need to understand the structure of the land.
William Smith's magnificent geological map of England and Wales, which was published
in 1815, was one early response to this need. Another was the formation of the Geological
Survey of Great Britain and Ireland, which was founded in 1832 and given official status
by an Act of Parliament in 1845. With full government backing, the professional surveyors
worked their way slowly up the country, starting in the south of England, mapping the
ground in great detail. This all took some time: the first Scottish sheet was not completed
until 1859, and the far north was not reached until the 1880s. What they found there was not
a region full of mineral wealth, but rather the most remarkable geology. It is a geology that
still attracts scientists from far and wide, and which has led to the establishment of what is
now the North West Highlands UNESCO Global Geopark. The journey to the understanding
of this particular geology was not a smooth one. In fact, it led to bitter divisions within
the geological community, and a debate that is known as the Highlands Controversy. It is
a drama that reveals as much about personalities and behaviour as it does about science.
This chapter follows these last visitors to the north-west, scientists both professional and

amateur, as they struggled to make sense of the complex landscape. It will make a fitting climax to all those prior journeys over this wild terrain, which Pennant aptly described in the eighteenth century as 'so torn and convulsed' (1998).

Even in the eighteenth century, the local inhabitants seemed to sense that the geology of the north-west was not straightforward. The Reverend William Mackenzie, in the 1799 *Statistical Account for Assynt* observed that 'all the track … to Layne [Lyne], being five English miles, abounds with limestone in different forms: but on the opposite side of the river Ah [Loanan?], there is not the smallest piece of limestone to be found. This severing of the high limestone rocks from the opposite mountainous bleak hills … can hardly be accounted for but by ascribing them to some uncommon convulsion of nature.' MacCulloch, too, was perplexed by the area. His *Description of the Western Isles of Scotland*, a work of great influence throughout the nineteenth century, deals mainly, as the title suggests, with the geology of the islands off the west coast. However, he did also include the far north-west mainland, and what he found there was most confusing. In an area of settled geology, one might expect to find layers of sedimentary rocks (that is, those laid down *in situ*, commonly in water, the fine particles gradually compressing into rock over millions of years), lying on top of one another in an ordered sequence, with younger rocks superimposed on older ones. In the north-west, MacCulloch found such sedimentary rocks – sandstone, quartzite, and limestone – but they were not in any obvious order. In some places – at Whiten Head for example – the quartzite was on top of the limestone, sometimes even with a second band of quartzite *below* the limestone, whilst in other areas it was the limestone that formed the uppermost layer.

The most settled geology was at the very edge of the west coast, where gneiss dominates, giving the land its familiar appearance; from the summit of any mountain in the region, you will look down on a landscape filled with small rocky hillocks, and innumerable lochs and lochans. Gneiss is an old rock (the oldest in Britain), what geologists term a 'metamorphic rock'. It perhaps started its life billions of years ago as a sedimentary type, but as a result of earth movements and further layers forming above, it was pushed down well below the earth's surface, where the pressure and heat changed, that is, metamorphosed, the composition of the rock into something much harder. Erosion above and further earth movements then brought this rock back to the surface.

Lying above the gneiss in places along the north-west coast, MacCulloch found an old sedimentary rock, sandstone of which are formed many of the mountains – Suilven, Stac Pollaidh – that rise so distinctly over the area. Sometimes, as on Quinag, for example, the sandstone is topped by a layer of white quartzite. However, MacCulloch noticed that even this was not consistent, for on Arkle, the quartzite rests directly on gneiss, not sandstone. Matters became more complicated still a little further east: he found areas in which the old,

"Quinag from the North. Red Sandstone resting on Gneiss, and capped unconformably by White Quartzite". A sketch by B.N. Peach, taken from The Geology of Sutherland *by H.M. Cadell.*

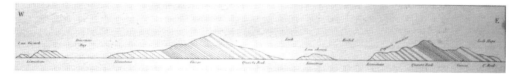

A section, taken from Volume 3 of MacCulloch's Description of the Western Islands of Scotland, *1819. It shows the rocks from Durness to Ben Hope, with layers of gneiss (metamorphic) apparently lying on top of limestone and quartzite (here marked 'quartz rock') – both sedimentary.*

metamorphic rocks were lying *on top of* the sedimentary layers. That was very odd indeed. How did such an old rock that had spent many millions of years in the depths of the earth where it had undergone chemical changes, end up on top of unmetamorphosed rocks which were much more recent?

MacCulloch found few fossils in the rocks of the north-west. This was unfortunate as in the early days of geology, it was chiefly fossils that enabled scientists to date the various bands of rock. He did, however, find what seemed to be organic remains in some sections of the quartzite, which he referred to as 'imbedded cylindrical bodies' (MacCulloch, 1819). It is now known as pipe rock, and the remains are thought to be old worm burrows. This was an important observation: the pipe rock would eventually help in unravelling the overall structure, but at this stage, MacCulloch could offer no solutions, only pose questions.

Another important visitor, Robert Hay Cunningham, spent three seasons in the north-west between 1838 and 1841. He noted that 'presenting even now in its comparatively improved state, a difficulty of access, which is not inconsiderable to the traveller among its mountain fastnesses, Sutherland has hitherto been but partially explored' (Cunningham, 1841). His *Geognostical Account of the County of Sutherland*, which was published in 1841, is of particular significance. It won him esteem and prizes. He married in 1842, but died shortly afterwards at his home in Edinburgh, which was a tragedy not only for his young wife, but also for geology, for his work was of the highest calibre. Coming from a wealthy background, he had time in his short life to survey the geology of six mainland counties. That of Sutherland is his outstanding achievement. His account includes the first geological map of the county which shows he had a clear understanding of the layout of the various rock types over the entire area. He found another two narrow but significant bands within the sedimentary layers, one in the limestone containing small conical shells called salterella, which give the name to this type: salterella grit. The other was a distinctive brown colour, a more earthy band that contained flattened tubes that were interpreted as fossil seaweeds, but in fact are now thought to be, like pipe rock, the work of burrowing animals. These are the fucoid beds. Again, both these layers were to prove important in finding the solution to the structure. Cunningham's account only added to the confusion in the area. He realised that in the east, there was a second metamorphic rock in addition to the gneiss, which he described as mica schist. At Eriboll, the gneiss was on top of the quartzite, while east of Arkle, the schist was on top of the gneiss. So now there were two metamorphic rocks lying on top of the sedimentary layers.

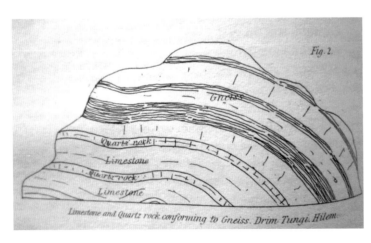

A section, from Cunningham's survey, showing gneiss sitting on top of layers of quartzite and limestone at Heilem, Loch Eriboll.

One more observer, an amateur geologist called Daniel Sharpe, deserves a mention at this point. He must have spent some time in the north-west before composing his paper *On the Arrangement of the Foliation and Cleavage of the Rocks of the North of Scotland* which was published in 1852. He spoke for many of his colleagues when he wrote that, on entering 'any district of gneiss or schist, in search of order … his first impression is one of despair' (Sharpe, 1852). The order he was referring to concerned the stripy bands one can see within individual rocks. Were they bedding planes, laid down when the deposits were forming in the rivers, or something more complicated? His suggestion that some of these bands might be the result of pressure *from the side* is of particular interest, for this was the first time that such an idea had been mooted, and it would not be the last (see colour section, page xiv).

These were the pioneer geologists who came into the area in the first half of the nineteenth century. In fact, another pair had ventured into the region in 1827, though their main interest was the sandstone that is found in the cliffs of Caithness. However, their enquiries took them as far west as Loch Eriboll, where like MacCulloch before them, they saw the quartzite below the limestone and the gneiss. They even wondered if this made the quartzite the oldest rock in the series. The senior geologist of this pair was Adam Sedgwick, who since 1818 had been Professor of Geology at Cambridge. His companion on this expedition was Roderick Impey Murchison, a man who, at the age of 35, had been involved in the science for only three years. As his biographer, Archibald Geikie, has pointed out, this was a time in geology when 'great work could be done by a man with a quick eye, a good judgement, a clear notion of what had already been accomplished, and a stout pair of legs' (1875). In fact, no one did greater work in nineteenth-century geology; knighted in 1846, his influence in this field was enormous. He

Left: A photograph of Professor Adam Sedgwick, by Kilburn of Regent Street, London.

Far left: Sir Roderick Impey Murchison, a photograph by C. Silvy, Bayswater, London.

would dominate the scientific establishment for much of his working life both as President of the Royal Geographical Society, and as Director General of the British Geological Survey.

This was not a bad record for a man who had shown no interest in science until he was twenty years old. In fact, he showed little academic prowess at school, joined the army as soon as he could, and went off to Spain under the Duke of Wellington at the age of 16. He boasted about a skirmish with the French at Vimieira, but made less of the bedraggled retreat to Corunna, from where he returned to England. Thereafter, he seemed destined for the life of a minor country gentleman, with a bit of hunting providing the only excitement. However, he joined the Royal Institution in 1812, and the contacts he made there, together with the lectures he attended, seemed to fire an interest in scientific life that found fulfilment in 1824, when he decided to devote his time to the science of geology.

He would have been delighted to have been in Scotland in 1827, first because he worshipped Sedgwick ('one of the kindliest, wittiest, merriest of companions' [Geikie, 1875]), and secondly because he felt a huge attachment to the north. At one time his family had been hereditary castellans of Eilean Donan Castle (see colour section, page xiv), one of the most photographed Scottish sites. One of his ancestors, Donald Murchison, gained distinction for resisting English rule after the defeat of the Rebels in 1715, by ensuring that the rents on the Seaforth Estate went to the banished Earl and not to the British Government. Sir Roderick clearly associated himself with this ancestor, for not only did he later erect a monument to Donald on the shores of Loch Duich, but he also commissioned from Edwin Landseer a painting depicting the scene, titled 'Rent-day in the Wilderness', which can be seen hanging in the National Gallery of Scotland. The artist used Roderick as the model for Donald, and on the rug in front of him there lies a snuff box, a precious heirloom, for family tradition has it that it was a gift from James the Pretender. Modern research doubts this, but it certainly came from a Jacobite sympathiser: on the top it is inscribed with the motto 'James Rex, Forward and Spare Not'.

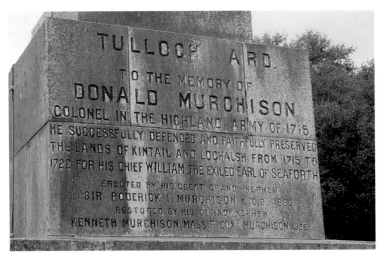

Murchison's monument to his ancestor, Donald, which stands on the shores of Loch Duich.

This was, perhaps, a suitable motto for Sir Roderick himself, for the 1840s saw him sweeping around Europe – first, Germany, then Russia. Sweep is the word: pulled by up to five or six horses, he would sit back in his calèche or tarantass, and assess the landscape, for that is how he liked to do his geology. 'With his faculty for quickly seizing the salient features of the geological structure of a country, he liked well to move swiftly from point to point, eye and note-book busy all the while noting and recording each point as he went along' (Geikie, 1875).

However, his greatest geological accomplishment had required more careful and methodical work. In 1831, he spent time in the Welsh borders, studying the rocks which lay below the sandstone layer on the surface. Using the fossil evidence, he was able work out the age and order of these rocks, and in so doing, created a whole new geological period which he named 'Silurian' after the tribe, the Silures, who had lived there under King Caractacus. He gave the simple title *Siluria* to the important book he wrote on the subject. It was first published in 1838 and, significantly, in view of the coming controversy, Geikie says the 'domain of "Siluria" became in his eyes a kind of personal property, over which he watched with solicitude' (ibid.).

There was certainly something rather imperious about the man. After his success in Russia, where he found large expanses of rocks dating from the Silurian period, 'he now felt himself entitled to assume the authority of a general of division. To many men who do not know him, this tendency assumed an air of arrogance' (ibid.). He loved to be called

The Illustrated London News *recorded Murchison's visit to Dudley in 1849. A figure can be seen addressing the huge crowd using a speaking trumpet.*

the 'Duke or King of Siluria'. William Conybeare, a fellow geologist, greeted him on his triumphal return from Russia as 'Dear and most illustrious Count Silurowski Ouralowski' (ibid.). In the public arena, there was a rather telling occasion of a gathering at Dudley, near Birmingham, where Murchison gave a lecture on the local Silurian rocks. The vote of thanks was given by the Bishop of Oxford, who, according to H. Woodward, took a gigantic speaking trumpet and addressed the crowd, 'giving one word at a time, to enable those present (several thousands, it is believed) to repeat it all together HAIL – KING – OF – SILURIA ... the vast assembly thrice responded with stentorian voices and most hearty hurrahs, and ever afterwards Sir Roderick was proud to be acknowledged "King of Siluria".' (Woodward, 1908).

It was perhaps in the guise of King of Siluria that Murchison sped north again, in August 1855. A Caithness man, Charles Peach, had discovered some fossils (not many, and of poor quality) in the limestone of Durness. It was possible that they could help to indicate the age of the rock. Peach knew that Murchison would be interested, and indeed he was. Since his work in Wales in the 1830s, rocks of the Silurian period had been found elsewhere. The French geologist Joachim Barrande spent ten years finding such rocks in Bohemia. In Britain, the very man Murchison had invited to join him on this present expedition north, James Nicol, had suggested in 1848 that those of the Southern Uplands in Scotland might be from this period. If the limestone in Durness could be dated to the early Silurian period, then all the layers above, including the metamorphic gneiss and schist might be later Silurian. This idea opened up the possibility of extending Murchison's so called 'Silurian Empire' over a vast new area, for the Moine schists, which Cunningham had recognised, reach a long way south into Inverness-shire.

Left: James Nicol, a photograph from Aurora Borealis Academica, *1899.*

Far left: Charles W. Peach, from Samuel Smiles's biography of Robert Dick, 1848. Peach was the father of Ben, who was to leave his mark distinctly on the Highlands Controversy at the end of the century.

In 1853, with the support of Murchison, James Nicol had been appointed to the Chair of Natural History at Aberdeen University, which meant he was well placed to explore these northern regions. He was acquainted with the uncertainties exposed by MacCulloch and Cunningham, having mentioned them as early as 1844 in his *Guide to the Geology of Scotland*. Judging by the fond biography found in *Aurora Borealis Academica*, a book subtitled *Aberdeen University Appreciations 1860 – 1899*, he was a man whose appearance and manners were in marked contrast to those of his imperious senior. Nicol was described as being 'On the shady side of middle life, somewhat large of bone, spare of flesh yet not lean, erect in figure and firm in gait … a man in hard condition, unused to luxury and capable of physical endurance … A kindly man, withal, to look at; but something in the firm straight mouth told of a possible dourness it were better not to provoke.'

He had no interest in his outward appearance. 'Sometimes his waistcoat would be buttoned awry, and then a whispered query would pass round the class whether the mistake should be geologically described as a "fault" or a "slip".' He was a kind man, and a lenient one when it came to exams. 'No student with ability and application sufficient to have brought him successfully to his magistrand year needed to fear that the professor of natural history would stand between him and the coveted degree.' However, he did once fail a student, a rather talented one at that: 'the professoriate pleaded for pity on the unhappy wight, but Nicol stood firm: "He said that the cow had no anal opening, and I [could not] pass him"' (*Aurora*, 1899).

Towards the end of his life, Nicol's voice degenerated into a husky falsetto, caused by what Bannerman, the author of the Appreciation, darkly refers to as 'the fatal underlying malady', though it did not prevent his students from mimicking him. Like the man, the language he used was unaffected, and he preferred plain English to scientific jargon where possible. He could be poetic too. No one who described Loch Maree in the following words could be accused of being a dry academic: 'A great fault, or rather a complication of faults, has formed the noble loch with its thirty islands, and has shivered the rocks into wild weather-worn peaks, divided by dark glens and rugged corries, down which the rivers hurry over linns and rapids. In this region the poet, artist, and tourist love to linger, and the geologist may spend weeks in studying the most wondrous sections in the British islands' (Nicol, 1866). Indeed, Loch Maree became one of the key areas in the controversy that was to unfold.

Nicol and Murchison would have been excited at the thought of the possible discoveries that lay ahead, but their trip was hampered by poor weather. Still, they covered a fair amount of ground: from Eriboll to Tongue, thence to Brora, and back across to Torridon. At Eriboll, they enjoyed the company of 'the excellent and hospitable Alex. Clark(e)' (Murchison Journal, 1857/9) whose uncle, Charles, had entertained William Daniell at Glendhu.

Murchison was keen to make the structure of the rocks fit his notion of extending his Silurian empire. He began to propose that the various layers to the east – limestone, quartzite, gneiss, and mica schist – had been laid down one after the other *in situ* (what geologists call a conformable sequence). Somehow the metamorphic layers had undergone their change at a later date. These were early days for structural geology, and anything was possible; processes might be found to allow such an event to have taken place.

It is interesting to see how, in the admirable search for scientific truth, ideas can be twisted for the sake of personal power and aggrandizement. Murchison was a proud Scot: proud of his lineage and of his links to Ross-shire. His monument to Donald Murchison was about to be erected on the shores of Loch Duich. There is little doubt that he seized on the possibility of extending 'his' Silurian Empire into the land of his ancestors, and allowed the idea to cloud his reason. The problem for Murchison was that his companion was not of the same mind. Both MacCulloch and Cunningham had described a rather more complicated picture, with no discernible structure, and for Nicol, the recent trip had clarified nothing. He and Murchison could not even agree on whether the quartzite lay above or below the sandstone.

Murchison, a caricature from Vanity Fair *magazine, 1870. He is described at the bottom as 'A Faithful Friend...' Nicol and Sedgwick discovered otherwise.*

The battle lines for the coming debate were being drawn up. For Murchison, the area showed predominantly a simple conformable arrangement: the layers of quartzite, limestone, quartzite (for he had to include two bands of quartzite), gneiss, and schist were laid down in the order in which they were seen. For Nicol, this was far too simple a picture. He spent further time on his own in the region in 1856, after which he presented a paper in which he suggested that the eastern gneiss and schist had somehow been 'forced over the quartzite' – what Oldroyd, in his classic account of the Controversy, points to as the 'first hint of thrust faulting in northwest Scotland' (Oldroyd, 1990).

Clearly, Murchison did not react well. In May 1856, Nicol wrote to him 'You and I seem to be gradually dropping out of acquaintance.'

Murchison, meanwhile, was also 'dropping out of acquaintance' with his idol, Adam Sedgwick. The situation was not dissimilar. With regard to rocks in Wales, Sedgwick felt that Murchison was trying to bring into the Silurian period rocks that actually belonged to an earlier period which Sedgwick had proposed, and which is now known as the Cambrian period. Murchison wrote to him assuring him that their friendship was akin to 'Siamese

A photograph by A. Brothers, Manchester. It is thought to be a meeting of the British Association for the Advancement of Science in Manchester, 1861. Various distinguished scientists are present, including Murchison, 2nd on the left. Perhaps not surprisingly, given their spat, Sedgwick can be seen well away from Sir Roderick, seated 4th from the right.

twins', but, on hearing nothing from his friend after he had sent him a copy of the new edition of *Siluria* in 1859 (perhaps not the most tactful thing to do under the circumstances), he wrote again 'clinging to the hope that the only bitter sorrow I have experienced in my scientific life may pass away, and that your old friendly feelings towards me may return' (letter, January 1859, quoted in Geikie, 1875).

Murchison made two more trips to the far north, one in 1858 with Charles Peach, and another in 1859 with Andrew Ramsay, who at the time was the Survey's second-in-command, and who replaced Murchison as Director General on the latter's retirement in 1871. There is a suggestion that both these companions had an interest in supporting the Murchison paradigm, and certainly neither offered any objections at the end of their expeditions in the way that Nicol had. Murchison, who was by now 66, was still full of energy. Geikie confirms that 'his powers of walking, though now of course manifestly on the wane, were still equal to the accomplishment of a ten or twelve mile tramp' (Geikie, 1875). His method had not changed: as he strode around 'his eye still kept its wonderful quickness in detecting the really salient features of the geological structures of a district' (ibid.). However, his notes reveal at least one occasion when he may have taken his eye off the ball. While Ramsay was on Quinag, Murchison met 'a sprightly, handsome girl, Isabella Fraser ... [who] offered to be my guide ... she was one of the handsomest Highland lassies I have seen, and if Edwin Landseer had been with me, he would have had her in one of his foregrounds.' At that very moment, they were walking along a key area of thrust planes, but Murchison had eyes only for Isabella, who, on seeing the swallow holes in the limestone, 'crept into one of them, and bounded out like a kitten' (Murchison Journal, quoted in Oldroyd, 1990).

Andrew Ramsay as a young man, from a photograph by George Wallich.

One can understand how welcome such distractions must have been during the long, lonely hours of surveying, but also wonder how hard he was really looking. Certain features in the area offer clear evidence of major geological turmoil. One such is the splendid rock face at the southern end of Loch Eriboll, called Creag na Faoillin, which had caught Cunningham's eye, and which provides signs to geologists of 'thrusts and overfolding (which) may be observed from a distance' (Oldroyd, 1990). Perhaps, as Oldroyd suggests, Murchison may not even have ventured down there, preferring to take the ferry across Loch Eriboll. Either way, these two trips to the north simply confirmed to him the correctness of his model, and it was in this positive state of mind that he and Ramsay attended the conference of the British Association for the Advancement of Science in September 1859. That year it was held in Aberdeen, Nicol's home ground, but it was Murchison who

"Loch Eribol, from Hailaim [sic] Inn". The Rev. C. Lesingham Smith took a tour in the far north in 1836, and provided this rare image of Loch Eriboll from Ard Neikie. The view is looking to the south, and Creag na Faoillin is the hill in the centre middle distance.

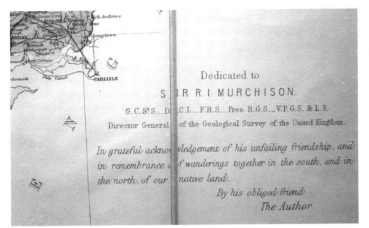

The dedication on Nicol's *1858* Geological Map of Scotland.

showed that he was in control. Murchison gave his paper to a large audience who greeted it warmly. Interestingly, Sedgwick leapt to his feet in support, even though he had not been in the Highlands since their trip together in 1827. Lyell gave the vote of thanks, referring to Murchison as 'the master of the Silurian', and the favourable reports in the local newspapers show that, even in Nicol's home town, it was the Director General who knew how to handle the press. The *Aberdeen Journal* recorded Sir Roderick's absolute certainty concerning the order of succession of the rocks – he had never seen 'a clearer order' (*Aberdeen Journal*, Sept. 1859). As if to ram home his advantage, the third edition of *Siluria* carried as a frontispiece an image, not from Wales, but from the far north-west of Scotland, an engraving of Quinag, viewed from Inchnadamph and Loch Assynt (see colour section, page xv). The Murchison paradigm was beginning to be firmly established in the public arena.

In the meantime, Nicol stood his ground, though he found himself increasingly isolated. He published a fine coloured geological map of Scotland in 1858, in which he equated the eastern gneiss with that found on the west coast, stating categorically in the notes that 'I am disposed to regard the great mass of this rock, extending from the north coast of Sutherland southwards through Ross-shire and Inverness-shire rather as belonging to an older period' (Explanatory Notes, Nicol, 1858). He thereby denied Murchison any Silurian territory. At the same time, the map bears the dedication: 'Sir R.I. Murchison … Director General of the Geological Survey of the United Kingdom. In grateful acknowledgement of his unfailing friendship, and in remembrance of wanderings together in the south, and in the north, of our native land.'

Mixed messages, perhaps, but whereas for Nicol a healthy disagreement over geological matters did not affect a partnership, Murchison's friendship was far from unfailing.

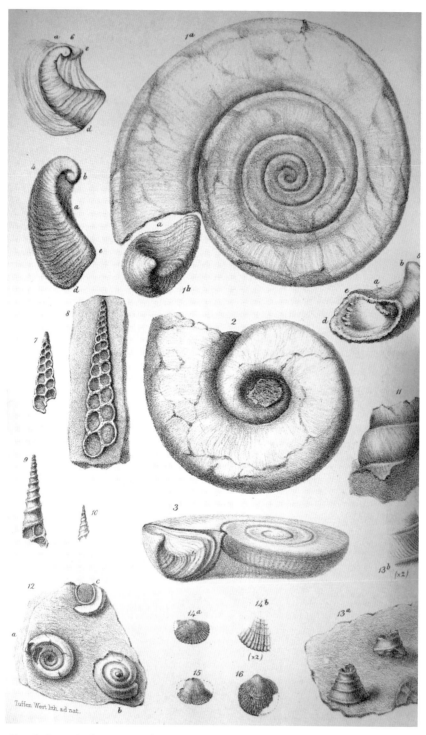

Fossils from the limestone of Durness. Murchison included these sketches in his article for the Geological Society of London Journal, *1859.*

In 1859, Murchison wrote an article in the Journal of the Geological Society of London, explaining his thoughts on the succession of rocks in the north-west. He included a map on which appears, for the first time, his vision of rocks of the Lower Silurian period dominating much of the north of Scotland (see colour sectio, page xv). He also provided sketches of the fossils that had been found at Durness by Charles Peach. This was his first public response to Nicol's map of 1858.

Sir Roderick made one more significant journey to the far north, and this time his choice of companion was surprising – Archibald Geikie, who, in 1860, was an ambitious 25-year-old employed in the Scottish branch of the Survey. He was a figure who achieved much eminence, and who became another important participant as the controversy unfolded. It is not easy to get a hold on Geikie's character. He was a most accomplished author, and with all his geological work, clearly a man of great energy and application. Edward Greenly, a geologist who worked on the Highland Survey from 1889, added that 'indefatigable worker though he was, Geikie had a winning trait: he was never "too busy"' (1938). Greenly himself wrote an autobiographical reminiscence in 1938, which is in part a slightly sentimental eulogy to his wife, but it also contains fascinating insights into the work of the Survey in the north. His assessment of Geikie is likely to be very fair. 'He was masterful, he could be a little chilly, and these traits tended to alienate men's sympathy. But we must not allow them to make us unjust. And he had another side: he could be very kind … There was something about him which, to say the very least, did not invite opposition from the members of his Staff … Though somewhat chilly in demeanour, he was excellent company … I was struck by his quickness on any new and difficult point, and even more with the beauty of his drawing' (Greenly, 1938).

Being an ambitious man, Geikie knew how to climb the career ladder, and what he wanted to achieve. He understood Murchison instinctively, and supported his views in the Highlands Controversy right up to the last moment; yet he was also able to stand up to him. Geikie's published works range from those dealing with serious geological matters, to delightful recollections, such as his *Scottish Reminiscences,* which abound with stories of his profession. One of his favourites was the anecdote which told of the geologist who, out in the field, sends a lad, for a small remuneration, back to the inn with a bag full of precious specimens. The lad, once out of sight, to ease his burden, empties the bag, and refills it later with stones collected from the vicinity of the Inn! You can sense that Geikie would indeed have been 'excellent company', but the picture of him as the convivial storyteller, chuckling over such stock-in-trade stories stands in contrast to that other aspect of his character, the schemer, working to get the new Chair at Edinburgh University, and eventually the Directorship of the Geological Survey. The two characteristics combine to form the man who was to dominate the science of geology in Britain during the later years of the Victorian era.

It is probable that Murchison saw in Geikie a fellow-spirit; someone he could understand. In addition, Oldroyd suggests that he may have been wanting to get the younger generation to accept his way of thinking concerning the Controversy, and saw the young Geikie as a means by which to achieve this. Geikie for his part wrote of Murchison's 'paternal kindness', and he must have relished the opportunities that this unexpected call to the great man's side offered. He certainly applied himself on the expedition, walking 32 miles in one day across the most rugged terrain between Kinlochewe and Ullapool, and following this feat with 20 miles to the north the next day. This latter path took him parallel to what was to become known as the Moine Thrust, but such was his haste that he seems to have missed it completely.

Murchison, meanwhile, was spending some of the time contacting his distant relatives in the area, expanding his knowledge of his ancestors. However, he and Geikie had met up at Loch Maree, and had puzzled over one particular rock they found in Glen Logan. It was one that had puzzled Nicol too. It was a rock that no one at that time could identify, and it would remain so for 20 years. It became known as the 'Logan Rock'.

By the time they headed further south, Geikie was utterly convinced that Murchison was correct. 'The general structure of the Highlands could no longer be considered doubtful … There was a fitness in the fact, that after wandering far and wide … he should return to his native Highlands and gather his last laurel from the rocks on which he was born' (Geikie, 1875). They had hoped to find evidence of the same succession further south in the lower Grampians, but in this they failed completely. Murchison observed poor Geikie 'cracking his brains and exhausting his energies in trying to coax these frightfully chaotic assemblages into the order of the north-west.' However, Sir Roderick 'stuck to his leading principle, from which no amount of contradictory detail would make him swerve' (ibid.).

"Kinlochewe, from Glen Logan". A misleading postcard published by Valentine in 1914. Glen Logan is in fact a pass leading off to the right of this image.

If nothing else, the tour cemented the partnership. They produced a geological map of Scotland together in 1861 – an expansion of Murchison's 1859 map – which proudly depicts a vast extent of Silurian territory. The map found its way, in revised form, into Geikie's *The Scenery of Scotland* (1865), a work of popular science that stated clearly Murchison's model. 'A few years ago these rocks presented a wild chaos of disorder … It was not until 1855 that Sir Roderick Murchison … established the true relations of the rocks of Sutherland and Ross, and obtained the key with which he has revealed the structure of the Scottish Highlands – a discovery the importance of which it is hardly possible to over-estimate.'

What with colleagues like Ramsay within the Survey on his side, and Geikie bringing on board the public and maybe a whole younger generation of geologists, it was as if all of Nicol's objections had been answered and the puzzle solved, bar a few minor details.

Nicol's last words on the subject are found in a small booklet, published in 1866, called *The Geology and Scenery of the North of Scotland*. The text consisted of two lectures he gave to the Philosophical Institution of Edinburgh. Aware that he was being shut out by the geological establishment, he added an appendix in which he answered some of the criticisms that Murchison had been aiming in his direction. One can sense the hurt he felt at his treatment. 'I must … express my most sincere regret that my illustrious opponent … has never found it convenient to meet me again in the North. I am convinced that we agree in so many of the essential points, that a few hours together in the field would bring us nearer in opinion than whole volumes of controversy.' Still, he hammered home his objections, principal of which was 'the impossibility of explaining how a great mass of strata, resting on unaltered sandstones and fossiliferous limestones can have been metamorphosed into gneiss, mica-slate and hornblende rocks.'

Murchison had no answer to this point, but he was not in a conciliatory mood as far as Nicol was concerned – 'I wish you would hammer him' (letter, January 1865, quoted

Geikie is one of the few geologists to appear on a cigarette card – Mitchell's Cigarettes, a series of "Famous Scots", 1933.

in Oldroyd, 1990), he wrote to Geikie. There was no need: his hold over the geological establishment ensured that his theories remained accepted and unchallenged for 18 years. During this time, the Survey was busy further south, while the north remained a remote and isolated area that few geologists, amateur or professional, thought to visit. Geikie, meanwhile, was making his way up the career ladder at some speed, with Murchison's help. The Director General obtained extra funding for the Geological Survey in 1866 which enabled him to split the English and Scottish departments into two separate organisations. Geikie was appointed Director of the Scottish office. Then, in 1871, following his donation of £6,000 to the University, Murchison's dream of a Chair of Geology at Edinburgh was fulfilled. The first Professor appointed was Archibald Geikie.

Murchison was showing little sign of slowing down. 'At night went to the annual ball at East Grinstead. Danced a quadrille with Mrs Mortimer West, and a tempête with the youngest Miss Stirling, Lady Caroline's lovely daughter, and a reel with Lady Arabella. Pretty well for a chicken of 1792, who had been geologizing all day' (Geikie, 1875). He made his last trip to Scotland in 1869, suffered a stroke in 1870, and died in 1871. Geikie describes his last visit to Sir Roderick in the biography. Murchison tried to speak, but could not. He then tried to write, 'but his fingers could no longer form any intelligible writing. His eyes filled with tears, and he sank back into his chair' (ibid.). At his funeral, the Queen sent her carriage as a sign of respect, and Gladstone was present among the mourners.

James Nicol died in April 1879. By this time, the establishment model for the structure of the rocks in the Highlands was once again under scrutiny. One hopes Nicol was aware of that before he died. He had received little encouragement in recent years, though in 1877 he met up with J.W. Judd, a geologist who had worked with the Survey in England between 1867 and 1871. Judd came to the Highlands expecting to support Murchison's views, but once there found himself in disagreement, and later was won round to Nicol's way of thinking on the matter. In view of the fact that Judd was soon to be embroiled in a rather bitter dispute with Geikie over details concerning the geology of Skye, it is not clear whether personal animosity played a part in this conversion, but Nicol must have been cheered at gaining a distinguished ally.

J.W. Judd, a portrait from the Illustrated London News.

The man who set the debate going again was, in fact, an amateur, a Welsh surgeon based in London called Henry Hicks. Amateurs played an increasingly important part in this controversy over the next few years. The science of geology had not yet progressed to such a degree of specialisation that their contribution was of no interest, yet as a group they began to feel the disdain of the professionals in the Survey. They may well have sensed a weakness in the establishment position regarding the rocks of the north-west, and, with access becoming ever easier, headed north to see what they could find. Hicks read his paper to the Geological Society of London in 1878. He raised no new points, but it inspired another amateur, Wilfred Hudleston, to review all the theories concerning the north-west geology, which showed that the matter was far from resolved, as the establishment would have it. The centre of attention was beginning to focus on Loch Maree. Hicks had reminded his listeners of the strange rock at Glen Logan, which prompted others to head for Ross-shire – Thomas Bonney, for example, a man at the cutting-edge of the new science of petrography. This discipline studied the composition of rocks, using equipment like the microscope, and it was fast replacing the need for fossils as a means of dating and categorising rocks. Hicks thought the Logan Rock was igneous, whereas to Bonney, it was gneiss. Hudleston simply observed that 'wherever there is Logan Rock, trouble is sure to ensue' (1882). He also noticed that wherever there was an upper quartzite band, there too was Logan Rock to be found.

Huddleston was joined at Loch Maree by another geologist, Matthew Forster Heddle, a wonderfully colourful character whose hill-walking feats earned him the distinction of becoming one of the first honorary members of the newly-formed Scottish Mountaineering Club. Heddle's expertise was as a mineralogist, and as such he was fascinated by the rock at Loch Maree – it was he who named it 'Logan Rock'. He said it was 'the only rock I ever examined in which I found no minerals' (*Mineralogical Magazine*, 1884). He published a fine geological map of Sutherland in 1882, marking on it the places where Logan Rock was to be found. He labelled it as 'igneous rock of the quartzite', but in fact he decided it was not igneous, but gneiss, somehow metamorphosed *in situ* (see colour section, page xvi).

Meanwhile, Geikie stuck firmly to the Murchison model, making three trips to the north-west between 1880 and 1881. On the first of these, he took with him Edward Hull, the director of the Irish branch of the Survey. It is perhaps no surprise to find that Hull came out firmly in favour of the establishment view, though Geikie may well have shielded him from the more controversial sites. It appears Geikie chose not to visit Loch Maree on any of these trips, even though it had by then become a key area in the debate. Sixty years had passed since MacCulloch first surveyed the region, and there was still no solution in sight. Clearly, some other approach was needed to come to an understanding of the structure of the area, and there were two men who realised what had to be done. Both of these men came from outside the Survey establishment.

Dr Charles Callaway was one of the first graduates in this country to hold a degree in geology. His professional career included curatorial posts at various institutions, including the New York State Museum in Albany. His interest was in the older rocks, and he visited the north-west region in 1880. At Durness he mapped small areas with great care, plotting the dip[1] of each of the rocks there. In the light of his findings, he thought it likely that neither the

---

1 Dip is defined as the steepest angle of descent of a tilted bed of rocks relative to a horizontal plane.

gneiss nor the schist was younger than the rocks over which they lay. Already, Murchison's long-held theories were starting to crumble, but Callaway's researches that year produced no major breakthrough, unlike those carried out in the summer months of 1881 and 1882.

Dr Callaway presented a paper to the Geological Society in May 1883, in which he set out the results of his research over the past three years. It was a defining moment in the history of the Controversy: not a resolution, but a huge leap forward. The progress had come about as a result of methodical, detailed mapping of specific areas within the region, some of which had already received much attention. Around Ullapool, for example, where Geikie had walked with such speed and confidence, and at the southern end of Loch Eriboll, on the rockface of Creag na Faoilinn, Callaway found all sorts of faulting and folding of the rocks, including remarkable S-shaped structures. He talked of rocks being crushed, and had found pipe rock at Glencoul, not far from Charles Clarke's house, which showed clear distortions suggesting huge lateral pressure coming in from the east. In the same site, he found the quartzite losing all trace of its 'classic' structure, passing into quartz-schist. He accounted for this by enormous pressure due to the quartzite being 'reflexed again and again in closely adpressed folds' (Peach and Horne, 1907). Nothing like this had been encountered before in British geology, and it was certainly a long, long way from the smooth succession favoured by the establishment. Worse was to follow for its paradigm, for Callaway utterly rejected the idea of an upper quartzite layer. By following on foot the outcrop on the hills of Drum na Teangra, he could trace the beds which revealed an S-shaped formation, suggesting overthrow – the two layers of upper and lower quartzite are actually formed by the one layer folded and laid over itself.

As Callaway delivered his findings, the Survey must have been listening with increasing anxiety. Seated within the hall was another individual who had a particular interest in what he was hearing: Charles Lapworth. The Survey had already come across this gentleman, and it had not emerged from the encounter with great credit. Lapworth spent the early part of his career as a schoolteacher, until he was offered the Chair of Geology and Mineralogy at Mason's College in Birmingham in 1881. His first post, from 1864, was at a school in Galashiels, a market town situated in the Southern Uplands to the south of Edinburgh. His time there happened to coincide with the Official Geological Survey of this region. They had found a coarse sandstone known as greywacke, some five miles thick, dominating the area. The Survey decided that this

Photograph of Charles Lapworth, from the History of the Geological Society, Glasgow, *published in 1908.*

rock had been laid down as they found it, a single ascending series. Lapworth spent a number of years researching the area in meticulous detail. He found fossils, called graptolites, within the sandstone, and was able to prove from their distribution that the area had undergone a remarkable amount of faulting which had resulted in the beds repeating themselves: the depth of the greywacke was nothing like as thick as the Survey had suggested, they had failed to spot any of the faulting, and they had later to return to re-survey the entire region.

Lapworth, according to Thomas Bonney, 'listened with great pleasure to the paper … Dr Callaway's interpretation of the structure of that area [Assynt] coincided with what he himself worked out in the Eriboll district.' He had spent some time in Sutherland in 1882, and probably formed an impression of the true relationship of the rocks. However, he published nothing until later in 1883.

Lapworth had arrived on the north coast in early August 1882. Oldroyd believes he came fully armed with the latest research from the Continent regarding mountain-building, particularly that of the French Alps. The terrain there is much younger than that of the northern Highlands, and contains much fossil evidence which helped considerably in the job of unravelling the structure. Continental geologists were considering not just vertical pressures from below the Earth's crust, but also lateral pressures. If rocks are squeezed from the side, they fold into an S-shape and eventually shear – the beginning of a thrust structure that can bring old rocks on top of newer ones.

Lapworth at first looked for fossil evidence which had helped him work out the structure in the Southern Uplands, but he found little to help him at Faraid Head, so he moved on to Loch Eriboll. His method involved mapping in meticulous detail, and in order to do this, he needed to understand the basic structure of the area. Looking at the rocks on the eastern shore of Loch Eriboll, he was able to find further categories within the basic rock types. Limestones, for example, were grey and mottled at the top of the series, dark and flaggy in the middle, and contained shells and salterella at the bottom. In the fucoid bed, itself a very distinctive earthy colour, he found an upper band of dark blue shale, while the quartzite he divided into four layers, with the pipe rock at the top.

Armed with this knowledge, he set to work on his mapping, and within a month, seems to have gained for himself a clear picture of the basic structure of the area and an understanding of how it had been formed. Using all the available clues from his sub-divisions, he could see not only which rock type was where, but also whether it was the right way up, turning over, or upside-down. The areas that interested him in particular were those at Ben Arnaboll and the deserted terrain out towards Whiten Head. This was exhausting work which, he told Bonney in a letter, he carried out night and day from his base in a shepherd's hut at Heilem.

Lapworth announced nothing at the end of 1882, and returned to the area in 1883. At the end of this visit, he felt able to publish his findings in the 1883–84 *Proceedings of the Geologist's Association.* John Horne records straightforwardly in the *Geological Survey Memoir* of 1907 that Lapworth 'believed that the Highlands of Scotland include a portion of an old mountain system, formed of a complex of rock-formations of very different geological ages, which have been crushed and crumpled together by excessive lateral pressure, locally inverted, profoundly dislocated, and partially metamorphosed'. Lapworth himself was rather more descriptive of this process:

> Conceive a vast rolling and crushing mill of irresistible power, but of locally varying intensity, acting not parallel with the bedding but obliquely thereto; and you can follow the several stages in imagination for yourself ... Shale, limestone, quartzite, granite, and the most intractable gneisses crumple up like putty in the terrible grip of this earth-engine – and all are finally flattened into thin sheets of uniform lamination and texture (Letter, Lapworth to Bonney, September 1882, quoted in Oldroyd, 1990).

These new sheets he called 'mylonite'. These were rocks with a stripy pattern – 'stripy as Regency wallpaper' Richard Fortey calls it in his book *The Hidden Landscape* – and it is found in the areas where the crushing and pulverizing was most intense. In the vicinity of Ben

Mylonite, at Loch na Maole, near Ullapool. A photograph by Chris McNeill.

Arnaboll, Lapworth found two major thrust planes: the Arnaboll Thrust which brought the ancient gneisses over the sedimentary rocks and the Moine Thrust which brought the Moine schists over from the east, for a distance that has been calculated to have been at least ten miles. In the process these rocks would have been further metamorphosed *in situ,* thus explaining the differences between the gneiss of the east and the west which Murchison had noted. Sir Roderick's upper quartzite layer, however, was eliminated in a most theatrical way when Lapworth met colleagues Teall and Blake out in the field, towards Whiten Head. Teall said that Lapworth placed him on the lower quartzite band, 'with strict instructions to walk along it, making sure I never left it. He and Blake took up their positions on the "Upper Quartzite", and moved in a direction roughly parallel with me with the same care. Progress was slow, for in such a disturbed area it was necessary to examine every inch of the ground. Finally we met on quartzite, shook hands, and declared that beyond doubt the "Upper Quartzite" was merely the "Lower Quartzite" brought up again by the disturbances of which we had already seen such striking evidence' (Teall and Watts, Lapworth Obituary, 1921, quoted in Oldroyd, 1990).

As for the bands that Murchison had thought were bedding planes, many of these are what modern geologists call thrust planes, along which the rocks were forced inexorably from the east towards the west. Lapworth describes them as, 'gliding-planes, along which the rocks have yielded to the irresistible pressure of the lateral Earth-creep during the process of mountain-making' (Lapworth, Geo. Assoc. Proceedings, 1883, quoted in Oldroyd, 1990).

Such was the intensity of the work, and perhaps the obsessive nature of Lapworth's character, that he suffered some sort of breakdown towards the end of the 1883 season, 'feeling', as Sir Edward Bailey described it, 'the great Moine Nappe grating over his body as he lay tossing on his bed at night' (Bailey, 1952, quoted in Oldroyd, 1990). He was forced to take six months off, leaving his colleagues to press the case for his model to the Survey, and in particular to Geikie. The Director General returned to the area in 1884, and after being led to the important sites by his own surveyors who had been in the area since 1883, he reluctantly realised the inadequacy of the Murchison paradigm to which he had adhered so loyally over the years. Geikie was not a man to show weakness, but Lapworth was told that 'his aspect after yielding seems to have been pitiable' (letter, January 1885, quoted in Oldroyd, 1990). Coming so soon after the Southern Uplands debacle, the ignominy for the Director General of the Survey must have been intense, and Greenly wrote that the side of his character that 'did not invite opposition from the members of his Staff' (1938) was much less pronounced after 1884. However, yield he did, though he was reluctant to give credit where it was due. Rather than publishing the news in the Survey's own literature, he issued a brief paper in the November issue of *Nature* magazine, stating clearly the new findings, and introducing a longer paper by his surveyors Benjamin Peach and John Horne. At no point did Geikie mention the names of Lapworth, Callaway, or those that came before, who had led the way to these discoveries, thereby implying that it was the Survey itself that had done all the work.

The amateurs were of course furious. Three years later, in the 1887 edition of the *Scenery of Scotland*, Geikie was just about able to include the name of Lapworth in the story, but it feels a little grudging. 'It was not until the year 1884 that the ground having been studied in great detail, conclusions were arrived at independently by the Geological Survey and Professor Lapworth, which have at last given the key to the problem.' To some extent this is true. Greenly is adamant that, when Lapworth met Peach and Horne out in the field, he did not disclose to them any of the details of his findings, though he might have indicated to them that the answers all lay in the Eriboll region. Geikie, whilst being perfectly willing to acknowledge his error in supporting the incorrect model, continued to work hard to exaggerate the contribution of the Survey to finding the explanation.

Perhaps that was his role as Director General. His stance was helped by the diligence of his team, which in itself has become the stuff of geological legend. A glance at a modern geological map of the area will show just how complicated the work was, and the surveyors rose to the task with exemplary professionalism and ability. The names Peach and Horne are to this day spoken of with awe in geological circles. A memorial to their work has been set up overlooking Loch Assynt and Quinag – much the same view as that which Murchison used as his frontispiece for the second edition of *Siluria* when trying to claim it for his 'empire' (see colour section, page xv). You will also find the pair waiting for you in a display at Knockan Crag, which is now rightly celebrated as one of the classic sites of the Moine Thrust itself. Press a button, and they will tell you of their exploits. How many geologists have enjoyed such celebrity status?

The memorial to the work of Ben Peach and John Horne, erected in 1980. It overlooks exactly the view chosen by Murchison for the frontispiece of his book Siluria. *Photograph by Chris McNeill.*

If the general structure of the area was now understood, there were still plenty of questions to be answered. What, for example, had caused these lateral forces? The theory of Continental Drift and Tectonic Plate Movement developed gradually throughout the twentieth century. It is the movement of these plates over the earth's surface that exerts a huge force from the side.

The detailed, methodical work of Callaway, Lapworth, and, later, of the Geological Survey had revealed an area of the most remarkable, even cataclysmic, geology. It was an old mountain system, the descendants of which now dominate this wild terrain: Suilven – the 'Sugar Loaf', Ben Loyal – the 'Queen of Scottish Mountains', Foinavon, Arkle, Stac Pollaidh, An Teallach, and many more, for this is one of the richest hill-walking areas in the British Isles. It is a region that has amazed visitors for centuries: 'so very extraordinary in its nature', according to General Roy, writing in the eighteenth century, 'we may suppose some hundred of the highest mountains split into many thousands of pieces, and the fragments scattered about' (1793). Charles St John thought Eriboll 'as wild as can well be imagined; consisting of irregular piles of gray[*sic*] rocks, thrown together in every kind of confusion' (1884). Nicol understood that such confusion was an integral part of the geology of the region. Without

The fame of the area soon entered the public domain. Here are two photographs of the same view, near Inchnadamph. The image on the left is by the professional photographer George Washington Wilson. On the right is an image from an album of modest holiday snaps dated 1908. After visiting Ben Loyal and Loch Assynt, the photographer presents this shot as "Thrust Plane, Inchnadamff."

full knowledge of the process that caused it, he suggested that it was not surprising, 'that along this line, the strata should be crushed, contorted, thrown into apparently discordant positions [...] That, masses of one age should be brought into contact with masses of another, even widely different age [...] That wedges of the higher limestone or quartzite should lie alongside, or even lower down than the once deep-rooted Torridon sandstone or the gneiss [...]' (Nicol, 1866). In the face of such evidence for a cataclysmic event, it seems remarkable that Murchison, Geikie, and the geological establishment should cling for some 30 years to so simple an explanation, in which the conformable succession of rocks, as Geikie saw it, had 'proceeded according to definite laws, [which] in their beauty and symmetry, are still the mode in which the Creator regulates the economy of the world' (*North British Review*, 1861, quoted in Oldroyd, 1990). Such sentiments may have brought comfort at a time when Darwin's revelations were throwing science and religion into some confusion, but to maintain these theories in the face of the evidence required both a somewhat blinkered attitude and an aggressive stance. Such a position hardly embraced the spirit invoked by Francis Bacon's challenge to all scientists, quoted at the front of the *Geological Society of London Centenary Record* in 1907:

> But if any human being desire to push on to new discoveries instead of just retaining and using old; to win victories over Nature as a worker rather than over hostile critics as a disputant; to attain, in fact, to clear and demonstrative knowledge instead of attractive and probable theory; we invite him as a true son of Science to join our ranks, if he will, that, without lingering in the fore courts of Nature's temple, trodden already by the crowd, we may open at last for all the approach to Her inner shrine.
>
> (Francis Bacon, Novum Organum)

EPILOGUE

In the year 1907, the Geological Society of London celebrated its centenary in some style. Representatives were invited from all over the world, and the festivities lasted from 26 September to 3 October. Praise for the work of the Society over the previous 100 years flooded in – America, Mexico, Egypt, India and Argentina, amongst others, all sent their congratulations. The Russians remembered especially 'Sir Roderick Impey Murchison, the famous explorer and investigator of the greater part of Russia' (Geological Society Centenary Record, 1907).

Sir Archibald Geikie, KCB. A photograph taken from the booklet recording the Centenary Dinner of the Geological Society of London, 1907.

Various excursions were offered to those attending the celebrations: the Lakes, Lyme Regis, and Edinburgh. 'At the last moment, and at the urgent and unanimous request of those visitors who had given their names as joining the Edinburgh Excursion, it was agreed that this party should proceed to the North-West Highlands of Scotland instead.' Such was the fame of the area now, at least in geological circles. They visited Assynt and saw the Glencoul Thrust, and the Moine Thrust too at Knockan, where they 'to their great satisfaction, walked for a quarter of a mile on the bare plane of this Thrust, exposed in a small stream east of Knockan Cliff.' They even enjoyed good weather.

The Reception Committee for the centenary celebrations included many names of note – Lapworth, Judd, and Bonney amongst them. Heading it all, of course, was Sir Archibald Geikie, recently knighted, who had moved on from the Survey to be President of the Royal Society, as well as of the Geological Society of London. The record contains a photo of him. The pose is perhaps meant to convey dignity and seriousness, but is there a touch too of anxiety?

Ben Peach was not in the Reception Committee, but he was there at the dinner. So too, Greenly with his wife, whom he so adored. Lapworth was given a place on the high table, well to the right of Geikie, and Bonney was nearby, while Horne was well to Geikie's left. Judd, on the other hand, whose spat with Geikie had become quite nasty, was in the body of the assembled guests. Geikie could look straight down on him from his position on the high table. In all professions, occasions like this are fraught with tensions and politics that simmer barely below the surface, and it would have been intriguing to have observed how the various personalities interacted.

The year 1907 also saw the publication of the official *Memoir of the Geological Survey in the North West of Scotland*, now a much-prized volume. The chapters by John Horne give full credit to those that edged their way towards the truth, from MacCulloch, through Cunningham and Nicol, to Callaway and Lapworth. The full record of the Survey itself, under the magisterial eyes of Peach and Horne, is then revealed in detail. Nothing has done more to put this remote corner of the country on the map, both metaphorically and geologically.

I am standing overlooking Loch Eriboll. The deep water is calm as I look down towards the southern end. There, in the middle distance, is *Creag na Faoilinn*, the cliff-face which so puzzled Cunningham, and might have enlightened Murchison had he seen it. To my right is Heilem, with its little isthmus of Ard Neakie, from where perhaps Murchison took the ferry,

The pier on the shore facing Heilem on Loch Eriboll. The headland in the middle distance is Creag na Faoilinn.

and where Lapworth stayed while researching the area. Further to my right, in the distance, the cliffs lead to Cape Wrath, that far westerly point that was reached by intrepid travellers like Pococke, MacCulloch, and Daniell, but was not mapped by Pont. To my left are the 'Lapworth sections' overlooking the road that is now plied at speed by those doing the NC500. Beyond that stretches the Moine, over which the unfortunate Reverend Macdonald had to trudge before delivering his sermon at Tongue.

The sea is clear, an enticing blue so beloved of those Lowland painters, the Scottish Colourists. It is invitingly Mediterranean, but I know the water will bring a sharp gasp should I enter it. The air is fresh. A breeze, tinged with the smell of peaty earth, barely disturbs the water. The silence is unbroken. Beneath my feet I can sense the power of the landscape, almost as if the rocks were even now edging their way inexorably westward.

M! TROTTMAN EN ÉCOSSE.

N.°16.

Vue générale de l Ecosse .

BIBLIOGRAPHY

ARCHIVE SOURCES

Edinburgh, National Library of Scotland Manuscript 9233, Anon, from Dover. *Journal of a Few Days from Home.*
  1856. MSS.2507-8, Journals of James Robertson.
Geological Society Archives Murchison Journal, 1857–59.

PRINTED SOURCES

Alexander, William. *Notes and Sketches Illustrative of Northern Rural Life in the Eighteenth Century.* David Douglas, 1887.

Anderson, George, and Peter. *Guide to the Highlands and Islands of Scotland.* Tait, 1842. New Edition.

Anderson, James. *An Account of the State of the Hebrides and Western Coasts of Scotland.* G.G.J. and J. Robinson, 1785.

Aurora Borealis Academica: Aberdeen University Appreciations 1860 – 1889. Aberdeen: University Printers, 1899.

Barron, James. *The Northern Highlands in the Nineteenth Century.* Robert Carruthers & Sons, published in three
  volumes: 1903, 1907, and 1913.

Bonehill, John, and Daniels, Stephen. *Paul Sandby, Picturing Britain.* Royal Academy of Arts, 2009.

Boswell, James. *The Journal of a Tour to the Hebrides.* Penguin Classics, 1984. Edited by Peter Levi.

Boswell, James. *The Life of Samuel Johnson.* Isaac Pitman, 1907. Edited by Roger Ingpen.

Boué, Ami. *Ésquisse Géologique sur l'Écosse.* V. Courcier, 1820.

Bristed, John. *Anthroplanomenos; or A Pedestrian Tour through part of the Highlands of Scotland in 1801.* J. Wallis, 1803.

Brown, Peter Hume. *Early Travellers in Scotland.* David Douglas, 1891.

Browne, James. *Critical Examination of Dr. MacCulloch's Work on the Highlands and Western Isles of Scotland.* Daniel
  Lizars, 1825.

Burt, Edmund. *Letters from a Gentleman in the North of Scotland to his Friend in London.*
  Rest Fenner, 1818. 5th Edition.

Cadell, H.M. *The Geology of Sutherland.* The Sutherland Association, 1886.

Chambers, Robert. *A Biographical Dictionary of Eminent Scotsmen.* Blackie & Son, 1857.

Circuit Journeys of the Late Lord Cockburn. Douglas, 1889.

Clark, Alan. *The Last Diaries 1992–1999.* Phoenix, 2003.

Clarke, Reay. *Two Hundred Years of Farming in Sutherland.* The Islands Book Trust, 2014.

Combe, William. *The Tour of Doctor Prosody.* Matthew Iley, 1821.

Commissioners for the Roads and Bridges in the Highlands of Scotland. *Reports.*
  8 Reports, 'ordered to be printed' between 1804–1817.

5th Commissioners' Report, 1811

6th Commissioners' Report, 1813

7th Commissioners' Report, 1815

8th Commissioners' Report, 1817

Cooper, Derek. *Road to the Isles*. Routledge & Kegan Paul Ltd., 1979.

Cordiner, Charles. *Antiquities and Scenery of the North of Scotland*. London, 1780.

Cordiner, Charles. *Remarkable Ruins, and Romantic Prospects, of North Britain*. I. & J. Taylor, 1795.

Craven, James. *History of the Episcopal Church in Orkney, 1688–1882*. Peace, 1883.

Cunningham, Ian C. *The Nation Survey'd: Timothy Pont's Maps of Scotland*. Tuckwell Press, 2001.

Cunningham, Robert J.H. *Geognostical Account of the County of Sutherland*. William Blackwood & Sons, 1841.

Daniell, William. *A Voyage Round Great Britain undertaken between the years 1813 and 1823*. Tate Gallery, 1978. 2 volumes.

Defoe, Daniel. *Memoirs of a Cavalier*. Lister, 1750.

Dixon, John. *Gairloch in North-West Ross-shire*. Co-Operative Printing Co. Ltd., 1886.

Duncan, James. *The Scotch Itinerary, containing the Roads through Scotland on a New Plan*. James & Andrew Duncan, 1808.

Duncan, James. *Duncan's Itinerary of Scotland*. James Lumsden & Son, 1820.

Durie, Alastair J. (Editor). *Travels in Scotland, 1788 – 1881*. Scottish History Society, 2012.

Faujas de Saint Fond, B. *Voyage en Angleterre, en Écosse et aux Îles Hébrides*. Jansen, Paris, 1797. English edition Ridgway, 1799. Also Hugh Hopkins, 1907. (Revised edition of the English Translation, by Sir Archibald Geikie.)

Fortey, Richard. *The Hidden Landscape*. Cape, 1993.

Franck, Richard. *Northern Memoirs, Calculated for the Meridian of Scotland*. Constable & Co., 1821.

Fyfe, J.G. *Scottish Diaries and Memoirs, 1746 – 1843*. Eneas Mackay, 1942.

Garnett, Thomas. *Observations on a Tour through the Highlands and Part of the Western Isles of Scotland*. T. Cadell junior & W. Davies, 1800.

Geikie, Archibald. *The Life of Sir Roderick I. Murchison*. John Murray, 1875. 2 volumes.

Geikie, Archibald. *The Scenery of Scotland, Viewed in Connexion with its Physical Geology*. Macmillan & Co., 1st Edition 1865. 2nd Edition 1887.

Geikie, Archibald. *Scottish Reminiscences*. James MacLehose & Sons, 1904.

Geological Society Centenary Record, 1907. See Watts, W.W.

Geological Survey Memoir. *The Geological Structure of the North-West Highlands of Scotland*. HMSO, 1907. Benjamin Peach, John Horne, et al.

The Georgian Era: Voyagers and Travellers. Vizetelly, Branston and Company, 1834.

Gilpin, William. *Observations Relative Chiefly to Picturesque Beauty, made in the year 1776, on Several Parts of Great Britain, particularly the High-Lands of Scotland*. R. Blamire, 1789.

Gordon, Sir Robert of Gordonstoun. *A Genealogical History of the Earldom of Sutherland, from its Origin to the Year 1630*. George Ramsay & Co., 1813.

Grant, Elizabeth. *Memoirs of a Highland Lady*. John Murray, 1911. Edited by Lady Strachey.

Green, Samuel G. *Scottish Pictures Drawn with Pen and Pencil*. Religious Tract Society, 1886.

Greenly, Edward. *A Hand Through Time*. Thomas Murby, 1938. 2 volumes.

Gregor, Walter. *Echo of an Ancient Time*. J. Menzies & Co., 1874.

Grimble, Ian. *The World of Rob Donn*. Saltire Society, 1999.

Haldane, A.R.B. *The Drove Roads of Scotland*. David & Charles, 1973.

Haldane, Elizabeth. *The Scotland of our Fathers*. Alexander Maclehose & Co., 1933.

Haldane, James. *Journal of a Tour through the Northern Counties of Scotland and the Orkney Isles, in Autumn 1797*. J. Ritchie, 1798.

Hall, Reverend James. *Travels in Scotland by an Unusual Route*. J. Johnson, 1807. 2 volumes.

Henderson D.M., & Dickson J.H. (Editors). *A Naturalist in the Highlands: James Robertson – His Life and Travels in Scotland*. Scottish Academic Press, Ltd., 1994.

Hewitt, Rachel. *Map of a Nation. A Biography of the Ordnance Survey*. Granta Books, 2010.

Hogg, James. *A Tour in the Highlands in 1803*. Alexander Gardner, 1888.

Hogg, James. *The Three Perils of Man*. Canongate, 1996. Edited by Douglas Gifford.

Home, John. *Survey of Assynt*. T. & A. Constable Ltd. 1960.

Hudleston, Wilfred. *First Impressions of Assynt, 1882. Geological magazine*, Sept.

Jenkins, David, and Visocchi, Mark. *Mendelssohn in Scotland*. Chappell & Co., 1978.

Johnson, Samuel. *A Journey to the Western Islands of Scotland*. Penguin Classics, 1984. Edited by Peter Levi.

Johnston, Hamish. *Matthew Forster Heddle, Mineralogist and Mountaineer*. National Museums Scotland, 2015.

Keats, John. *Letters*. Gowans & Gray, 1901. Vols IV & V of the Complete Works.

Keats, John. *Letters, A Selection*. OUP, 1970. Edited by Robert Gittings.

Kemp, Daniel William. *Notes on Early Iron-Smelting in Sutherland*. Norman Macleod, 1887.

Knox, John. *A View of the British Empire, more especially Scotland*. J. Walter, etc. 1785. 2 volumes, 3rd Edition, greatly enlarged.

Knox, John. *A Tour through the Highlands of Scotland and the Hebride Isles in 1786*. J. Walter, etc., 1787. Modern edition, James Thin, 1975 (lacks Introduction, etc.).

Lawson, John Parker. *Descriptive Atlas of Scotland*. Edinburgh Printing and Publishing Co., 1842.

Lightfoot, Rev. John. *Flora Scotica*. R. Faulder, 1777.

Logan, James. *McIan's Costumes of the Clans of Scotland*. Bryce & Son, 1899. (Reduced size copy of the original).

Logan, James. *McIan's Highlanders at Home, or Gaelic Gatherings*. David Bryce and Son, 1900.

Logan, James. *The Scottish Gael; or Celtic Manners, as preserved among the Highlanders*. Smith, Elder & Co., 1831. 2 volumes.

Macleod, Anne. *From an Antique Land*. John Donald, 2012.

MacCulloch, John. *A Description of the Western Islands of Scotland, including The Isle of Man*. Constable & Co., 1819. 3 volumes.

MacCulloch, John. *The Highlands and Western Isles of Scotland, Containing Descriptions of their Scenery and Antiquities*. Longman, Hurst, Rees, 1824. 4 volumes.

Macgill, W. *Old Ross-shire and Scotland, as seen in the Tain and Balnagown Documents*. Northern Counties Newspaper and Printing and Publishing Co., Ltd., 1909. Supplementary volume published 1911.

McIan, R.R. *Clans of the Scottish Highlands*. Ackerman, 1845. Text by James Logan.

Mackenzie, Osgood. *A Hundred Years in the Highlands*. Geoffrey Bles, 1949.

Migglebrink, Joachim. 'The End of the Scots-Dutch Brigade'. In *Fighting for Identity. Scottish Military Experiences c.1550–1900*, 83–104. Edited by Steve Murdoch and Andrew Mackillop. Brill, 2002.

Mitchell, Ian R. *Scotland's Mountains Before the Mountaineers*. Luath Press, Ltd., 1998.

Mitchell, Joseph. *Reminiscences of my Life in the Highlands*. David & Charles, 1971. 2 volumes.

Moir, Donald G. (Editor). *The Early Maps of Scotland to 1850*. Royal Scottish Geographical Society, Volume 1, 1973. Volume 2, 1983.

Murchison, Roderick Impey. *Siluria*. John Murray, 1839, 1st Edition. With the Assynt frontispiece, from the 3rd Edition, 1859.

Murdoch, Steve, and Mackillop, A. (Editors). *Fighting for Identity; Scottish Military Experience, c 1550–1900*. Brill, 2003.

Murray, Sarah. *Useful Guide to the Beauties of Scotland*. George Nicol, 1799.

Necker de Saussure, Louis A. *Voyage en Écosse et aux Îles Hébrides*. Genève, J.J. Paschoud, 1821. English translation, Richard Phillips & Co., 1822.

New Statistical Account of Scotland. William Blackwood & Sons, 1845. Volume XIV (Inverness & Ross & Cromarty). Volume XV (Sutherland & Caithness).

Newte, Thomas. *Prospects and Observations; on a Tour in England and Scotland*. G.G.J. and J. Robinson, 1791.

Nicholson, William. *The Scottish Historical Library*. T. Whilde, 1702, 1st Edition; T. Evans & T. Becket, 1776. A New Edition.

Nicol, James. *Guide to the Geology of Scotland*. Oliver & Boyd, 1844.

Nicol, James. *The Geology and Scenery of the North of Scotland*. Oliver & Boyd, 1866.

Oldroyd, David R. *The Highlands Controversy*. University of Chicago Press, 1990.

Paton, Hugh. *Kay's Portraits*. Hugh Paton, 1838. 4 volumes.

Paxton, R., & Shipway, J. *Civil Engineering Heritage – Scotland Highlands and Islands*. Thomas Telford (publishers), 2007.

Peach, Benjamin, and Horne, John, et al. *The Geological Structure of the North-West Highlands of Scotland*. HMSO, 1907.

Pennant, Thomas. *A Tour in Scotland MDCCLXIX*. Birlinn, 2000.

Pennant, Thomas. *A Tour in Scotland, and Voyage to the Hebrides; MDCCLXXII*. Birlinn, 1998.

Pennant, Thomas. *The Literary Life of the Late Thomas Pennant*. Benjamin and John White, 1793.

The Penny Magazine. Society for the Diffusion of Useful Knowledge. Various, 1833–1837.

Playfair, James. *A Geographical and Statistical Description of Scotland*. Archibald Constable & Co., 1819.

Pococke, Richard. *Tours in Scotland 1747, 1750, 1760*. Scottish History Society, 1887. Edited by Daniel Kemp.

Portlock, J.E. *Memoir of the Life of Major-General Colby*. Seeley, Jackson, & Halliday. 1869.

Prebble, John. *Culloden*. Penguin, 1967.

Rackwitz, Martin. *Travels to Terra Incognita*. Waxmann Munster, 2007.

Reid, Stuart. *Scottish National Dress and Tartan*. Shire, 2013

Richards, Eric, and Clough, Monica. *Cromartie: Highland Life 1650–1914*. Aberdeen University Press, 1989.

Robertson, James. For his accounts, see Henderson, D.M. & Dickson, J.H.

Rolt, L.T.C. *Thomas Telford*. Longmans, 1958.

Roy, William. *The Military Antiquities of the Romans in Britain*. Society of Antiquaries, 1793.

Roy, William. *The Great Map. The Military Survey of Scotland 1747–1755*. Birlinn/British Library. 2007.

Sage, Donald. *Memorabilia Domestica: or Parish Life in the North of Scotland*. William Rae, 1899. 2nd Edition.

St John, Charles. *Tour in Sutherland*. David Douglas, 1884. 2nd Edition.

Scarfe, Norman (Editor). *To the Highlands in 1786: The Inquisitive Journey of a Young French Aristocrat*. Boydell Press, 2001.

Scotii Britannicus. *The Sporting Magazine*. July 1825. Article titled *Sporting Tour to Caithness*.

Scott, Sir Walter. *The Lady of the Lake*. John Ballantyne & Co, 1810.

Sharpe, Daniel. *On the Arrangement of the Foliation and Cleavage of the Rocks of the North of Scotland*. Philosophical Transactions of the Royal Society of London, 1852.

Shore, Charles John, Lord Teignmouth. *Sketches of the Coasts and Islands of Scotland*. John W. Parker, 1836. (Teignmouth was possibly the author of the anonymous contributions to the Penny Magazine – see above.)

Sinclair, C. *Scotland and the Scotch*. William Whyte & Co., 1840.

Sinclair, Sir John. *An Account of the Highland Society of London, from its Establishment in May 1788 to the Commencement of the Year 1813*. McMillan, 1813.

Sinclair, Sir John. *Statistical Account of Scotland*. Creech, 1791–1799.

Smart, J. *Ye Life and Troubles of ane Artist in Ye Highlands of Scotland*. John Davidson, No date.

Smith, Rev. C. Lesingham. *Excursions Through the Highlands and Isles of Scotland in 1835 and 1836*. Simpkin, Marshall, and Co., 1837.

Southey, Robert. *A Tour in Scotland in 1819*. Murray, 1929. Modern edition, James Thin, 1972.

Stewart, Colonel David. *Sketches of the Character, Manners, and Present State of the Highlanders of Scotland*. Constable & Co., 1822. 2nd Edition.

Stone, Geoffrey. *The Pont Manuscript Maps of Scotland: Sixteenth Century Origins of a Blaeu Atlas*. Map Collector Publications Ltd., 1989.

Taylor, William. *The Military Roads in Scotland*. David & Charles, 1976.

Trevor-Roper, Hugh. *The Invention of Scotland*. Yale University Press, 2014.

Turnbull, George. *Diaries of George Turnbull*. Scottish History Society, 1893. Edited by R. Paul.

Various. *The New Statistical Account of Scotland*. Edinburgh, 1845. 15 volumes.

Various. *The Saturday Magazine*. A number of issues, especially No. 104, 1834; 128, 1834; 139, 1834; 160, 1834; 214, 1835; 224, 1835.

Victoria, Queen. *More Leaves from the Journal of a Life in the Highlands*. Smith, Elder & Co., 1884.

Walford, Thomas. *The Scientific Tourist through England, Wales, and Scotland*. J. Booth, 1818. 2 volumes.

Walker, Carol. *Walking North with Keats*. Yale University Press, 1992.

Watts, W.W. *The Centenary of the Geological Society of London*. Longmans, Green, and Co., 1909.

Wills, Virginia (Editor). *Reports on the Annexed Estates 1755–1769*. HMSO, 1973.

Wilson, James. *A Voyage round the Coasts of Scotland and the Isles*. A. & C. Black, 1842.

Woodward, Henry. *The History of the Geological Society of London*. Longmans, Green & Co., 1908.

Wordsworth, Dorothy. *A Tour in Scotland in 1803*. James Thin, 1981.

Wraxall, Nathaniel. *Posthumous Memoirs*. Richard Bentley, 1836.

MAPS

Adair, John. *Nova Scotiae Tabula*, 1727. In Buchanan's *Rerum Scoticarum Historia*, 1727 edition.

Anderson, James. *A New Map of Scotland: The Hebrides and Western Coasts in Particular*, 1785.

Armstrong, Andrew and Mostyn. *A New Map of Ayrshire*, 1775.

Arrowsmith, Aaron. Each of the Reports of the Highland Road Commissioners is accompanied by a map by Arrowsmith showing the progress of the road construction 1807–1821.

Arrowsmith, Aaron. *Map of Scotland, Constructed from Original Materials*, 1807.

Blaeu, J. *Strathnavernia*; *Southerlandia*; *Extima Scotiae Septentrionalis*; *Scotiae Provinciae Mediterraneae*, and others, from the *Atlas Novus* 1654.

Boué, A. *Carte Geologique de l'Ecosse*, 1820.

Bryce, Alexander. *A Map of the North Coast of Britain*, 1744.

Buache, Philippe. *Carte minéralogique sur la nature du terrain d'une portion de l'Europe*, 1746.

Burnett & Scott. *Map of the County of Sutherland*, 1833.

Cowley, John. *A Display of the Coasting Lines of Six Several Maps of North Britain*, 1733.

Cunningham, R.J.H. *Geognostical Map of Sutherland*, 1841.

Dorret, James. *A General Map of Scotland and Islands thereto belonging, from New Surveys*, 1750.

Elphinstone, John. *A New and Correct Mercator's Map of North Britain*, 1745.

Faden, William. *Scotland, with its Islands, drawn from the Topographical Surveys of John Ainslie*, 1813.

Geikie, Archibald. *Geological Map of Scotland*. W. & A.K. Johnston, 1876.

Gordon, Robert. *Scotia Regnum*, 1654.

Heddle, M.F. *Geological and Mineralogical Map of Sutherland*. W. & A.K. Johnston, 1881.

Knox, John. *A Commercial Map of Scotland*, 1782.

Lizars, W.D. *General Map of the Caledonian Canal*, 1817.

MacCulloch, John. *A Geological Map of Scotland*. London, 1836. Later editions include Cruchley, 1843.

Mercator. *Scotiae Tabula III*. Atlas Minor, 1607.

Murchison, Sir Roderick Impey. *First Sketch of a Geological map of the North of Scotland*. Printed at The Geologist Office, 1859.

Murchison, R.I., and Geikie, Archibald. *First Sketch of a New Geological Map of Scotland*. W. and A.K. Johnston, and Blackwood, and Stanford, 1861.

Nicol, James. *Geological Map of Scotland from the most recent authorities and personal observations*. A. K. Johnston, 1858.

Playfair, James. *Geographical and Statistical Description of Scotland*. Constable, 1819.

Pont, Timothy. The Sketches of his tour of Scotland are in the National Library of Scotland.

Poracchi, Tommaso. *Scotia,* from *L'Isole Piu Famose del Mondo*, 1572.

Ramsay, A. *Geological map of Great Britain*, 1878.

Sharpe, Daniel. Map of North Scotland, in *On the Arrangement of the Foliation and Cleavage of the Rocks of the North of Scotland*, 1852.

Taylor, George, and Skinner, Andrew. *Survey and Maps of the Roads of North Britain or Scotland*, 1776.

Wallis, John. *A New Geographical Game, Exhibiting a Complete Tour through Scotland*, 1792.